Islam and Global Studies

Series Editors

Islam and Global Studies series provides a platform for the progression of knowledge through academic exchanges based on multidisciplinary socio-political theory that studies the human condition and human interaction from a global perspective. It publishes monographs and edited volumes that are multidisciplinary and theoretically grounded and that address, in particular, non-state actors, Islamic polity, social and international justice, democracy, geopolitics and global diplomacy. The focus is on the human condition and human interaction at large. Thus cross-national, cross-cultural, minority and identity studies compose the building block of this series; sub-areas of study to which Islamic theory and socio-political praxis can provide an alternative and critical lens of inquiry. It explores Islam in history and in the contemporary world through studies that:

a) provide comprehensive insights of the intellectual developments that have defined Islam and Muslim societies both in history and in the contemporary world;
b) delineate connections of pre-colonial Muslim experiences to their responses, adaptations and transformations toward modernity;
c) evaluate old paradigms and emerging trends that affect Muslims' experiences in terms of political state system, democracy, secularization, gender, radicalism, media portrayals, etc.;
d) show empirical cases of intra-Muslim and Muslim–Non-Muslim relations.

More information about this series at
https://link.springer.com/bookseries/16205

Abdur Rehman Cheema

The Role of Mosque in Building Resilient Communities

Widening Development Agendas

Abdur Rehman Cheema
Islamabad, Pakistan

ISSN 2524-7328 ISSN 2524-7336 (electronic)
Islam and Global Studies
ISBN 978-981-16-7599-7 ISBN 978-981-16-7600-0 (eBook)
https://doi.org/10.1007/978-981-16-7600-0

This Palgrave Macmillan imprint is published by the registered company Springer Nature Singapore Pte Ltd.
The registered company address is: 152 Beach Road, #21-01/04 Gateway East, Singapore 189721, Singapore

To my mother—may Allah she lives longer and healthier, and to the loving memory of my father, may Allah shower countless blessings on his soul.
Amen

Preface

What can I do for those who suffered in the aftermath of the 2005 earthquake in Pakistan? This was an inner call that I was motivated to respond to. In my generation's time, this was the deadliest event that moved me. I kept thinking about why this was to happen to those who believe in Allah and why the earthquake had to take thousands of lives? As I thought and studied deeper, I was to discover slowly and over time, why earthquakes happen and why earthquakes could not be blamed for widespread loss of life, property and assets. It was an extremely interesting journey of finding out the place and people where I was born and raised.

Though 15 years have passed since the 2005 earthquake, earthquakes continue to occur. Sadly, human beings continue to experience huge losses of lives, properties and assets due to these earthquakes, and more so in developing countries. States and especially developing countries lack institutional, human and financial capacity to effectively save lives and reduce disaster losses. Despite poverty and physical vulnerability due to geography, there is an increasing demand on communities to draw on their resources, build resilience and adapt to climate change. Easier said than done, there are multiple challenges to build resilient communities. Human perception, response and behaviour in the face of adversity, such as disasters and vulnerabilities, is one of the most significant difficulties. The human belief system and the institutions that shape this belief system have strong undercurrents on how we respond and recover from these disasters. This is what made me explore the role of the mosque, a community-based religious institution in the aftermath of the 2005 earthquake in Pakistan. There are around 0.25 million mosques in Pakistan, the sixth most populous country with a population of 220 million in the world.

When I was finalising this book, COVID-19 (Coronavirus), the pandemic hit the world in early 2020, still ongoing has claimed around 2.4 million lives. The virus has had unprecedented effects on the global economy and how social relationships are organised. Health authorities have enforced several protection measures, but the most pronounced ones relate to social distancing. Methods include quarantines; travel restrictions; and the closing of schools, religious places, workplaces, stadiums, theatres and shopping centres. Individuals are recommended to apply social distancing methods by staying at home, limiting travel, avoiding crowded areas, using no-contact greetings and physically distancing themselves from others. These measures are met with resistance everywhere though in diverse ways. State capacities overwhelmed, religious institutions and their leadership have been approached to help enforce social distancing.

As the world is grappling with the challenge of controlling the COVID-19 pandemic, this book offers critical insights on the significant but underutilised role of community-based religious institutions. It draws on the emerging evidence documenting diverse ways religious institutions are playing their role during the pandemic. If well planned and nuanced engagement was to replace ad hoc and occasional interaction, the book suggests that religious institutions could accelerate many times the achievement of the so-called material gains of ending poverty, hunger, inequality, malnutrition and gender imbalances, to name a few.

Religious institutions are playing an important and visible role in influencing the response of communities. Though this book is primarily about the role of the mosque in the aftermath of the 2005 earthquake and draws on my doctoral thesis titled "Exploring the Role of the Mosque in Dealing with Disasters: A Cases Study of the 2005 Earthquake in Pakistan" at Massey University, New Zealand, this also sheds light on the role of religious institutions in the ongoing COVID-19 pandemic. The book shows how religion and religious institutions are increasingly visible in the widening development canvas of human beings and not fading over time.

* * *

Islamabad, Pakistan

Abdur Rehman Cheema

Acknowledgements

Greatest thanks to my Lord, Allah Almighty, who created me as a human being and blessed me with all the bounties of life. I offer countless blessings of Allah on Prophet Muhammad (PBUH), the best of humans and a role model for humanity.

This book's journey started with a generous doctoral scholarship by the Higher Education Commission, Government of Pakistan in 2007. I appreciate this financial support from the Government of Pakistan. I am highly indebted to my supervisor, Professor Regina Scheyvens and co-supervisors, Professor Bruce Glavovic and Dr. Muhammad Imran who guided and supervised me during my doctorate at Massey University, New Zealand. Though substantially revised and updated, this book would not have been possible without incorporating some of the work that was part of my doctoral thesis titled "Exploring the Role of the Mosque in Dealing with Disasters: A Cases Study of the 2005 Earthquake in Pakistan" at Massey University, New Zealand.

I am also thankful to my mentor, Dr. Abid Mehmood working at the Sustainable Places Research Institute, School of Social Sciences, Cardiff University, UK and Charles Wallace Trust, Pakistan. This book would not have been possible without a six months' Chevening Fellowship at the Centre for Islamic Studies at the University of Oxford during 2019–2020. At the Centre, I am obliged to my colleagues namely Adeel Malik, Maulana Ibrahim Mohammad Amin, Michael Feener, Moin Ahmad Nizami, Muhammad Meki, Muhammad Talib, Nassef Manabilang

Adiong and Talal Al-Azem for their time, company and useful conversations. My colleague and friend, Dr. Faisal Abbas, has been a continuous source of inspiration and encouragement for me to accomplish this book and I am indebted to him. Last but not the least, I am also grateful to my colleague, Dr. Mehvish Riaz, who proofread the manuscript with great care and interest.

I am also immensely grateful to my loving family members for their encouragement and support to complete this work.

Islamabad
2021

Abdur Rehman Cheema

Praise for *The Role of Mosque in Building Resilient Communities*

"This book narrates untold stories about the complex nature of mosque, community organisations and development practices in Pakistan. Without doubt, this book further triggers passionate debates on the role of religious organisations in development practices. Abdur Rehman presents his decade long research on the role of religious institutions in disaster risk reduction in a thought-provoking monograph."

—Associate Professor Imran Muhammad, *School of People, Environment and Planning, Massey University, New Zealand*

"Dr. Abdur Rehman Cheema in this book has taken up a very rare subject which is essentially required by Muslim communities. Mosque, the most frequently attended religious place is packed with social energy. As visualized in the book, if this energy under the patronage of the Imam (Prayers leader) is correctly mobilized, it can help achieve success at every stage of disaster risk management cycle. The theme presented in the book can be effectively applied to inspire Muslim communities for disaster management."

—Brigadier Fiaz Hussain Shah, *Sitara-e-Imtiaz (Military), Retired, Former Director General Earthquake Reconstruction and Rehabilitation Authority (ERRA) and National Institute for Disaster Management (NIDM) /National Disaster Management Authority (NDMA), Pakistan*

"Dr. Cheema's outstanding scholarly work highlights the forgotten role of mosque as a catalyst for community development, social cohesion, and resilience building, and not merely as a place for observing rituals of faith."

—Tariq H. Cheema, MD, *Founder, World Congress of Muslim Philanthropists, Convener|Global Donors Forum, Chair of the Board|MuslimFunders, Chicago - London - Jeddah*

"This book from Abdur Rehman Cheema is an incredible endeavor to bring light to the potential future roles that the mosques and other religious institutions can play in disaster preparedness. These community-based religious institutions undertake an imperative part in the reaction, response, recuperation, relief, recovery, reconstruction, and preparedness phases of the disaster in nations with Muslim preponderance. The author confers different dimensions of the role of the mosques to resist, absorb, accommodate, adapt to, transform and recover from the effects of a hazard in a timely and efficient manner and passage magnanimity, pledge, and responsibility among the Pakistani community."

—Sajid Naeem, *Senior Program Manager/Country Representative Pakistan, Asian Disaster Preparedness Center (ADPC)*

"This sympathetic case study analysis of the role of mosques in rural Pakistan following the 2005 earthquake presents new insights into diverse dynamics of religion, economy, and gender in post-disaster contexts while highlighting both the possibilities and the limitations of humanitarian engagement with religious institutions in mitigating the impact of future disasters."

—Professor R. Michael Feener, *Center for Southeast Asian Studies, Kyoto University, Japan*

"This is an outstanding and much-needed contribution to help understand the role of faith and religiosity in humanitarian relief efforts and disaster risk reduction. Faith-based community set ups and organisations offer immense advantages in terms of access to areas where government machinery may struggle to reach. Abdur Rehman Cheema has brilliantly demonstrated how the mosque, as part of the civil society, faith-based institutions and networks can provide effective emergency response and recovery and improve community resilience."

—Dr. Abid Mehmood, *Senior Research Fellow, Sustainable Places Research Institute, School of Social Sciences, Cardiff University, UK https://www.cardiff.ac.uk/people/view/38131-mehmood-abid*

"For too long scholars have overlooked the significant role that religious institutions can and do play to assist communities all around the world in recovering from disasters. Dr Cheema's book makes important strides in this direction by focusing on the role of the mosque, particularly in terms of supporting the resilience of people in the face of a major earthquake in Pakistan."

—Prof. Regina Scheyvens, B.A. (Hons.), Ph.D., *Professor & Co-Director—Pacific Research and Policy Centre, Doctoral Mentor Supervisor, School of People, Environment and Planning, https://www.massey.ac.nz/massey/expertise/profile.cfm?stref=700330*

"Focusing on the role of the community mosque in disaster relief in Pakistan, this book offers a rare account of a community institution of importance to Muslims across the world. Based on empirical research, the book is an timely contribution to the debate about what disaster response and preparedness can be. The book is a must-read for anyone wanting to understand the role of religious institutions in disaster response in Pakistan and beyond."

—Dr. Kaja Borchgrevink, *Senior Researcher, Peace Research Institute Oslo (PRIO), Norway*

"The masjid/mosque is an integral part of Muslim society across globe. This book on tackling the climatic change crises by leveraging mosques in Pakistan or community-based religious institutions elsewhere is an outstanding and unique contribution. More than ever, the world desperately needs such indigenous community institutions to build adaptative and resilient societies in our fast-changing Anthropocene."

—Faisal Abbas, Ph.D., *Associate Professor (Economics), School of Social Sciences and Humanities (S3H), National University of Sciences and Technology (NUST), Islamabad, Pakistan*

Contents

Abbreviations

AJK	Azad Jammu and Kashmir
BBC	British Broadcasting Corporation
CBDRM	Community-Based Disaster Risk Management
DCO	District Coordination Officer
DDMA	District Disaster Management Authority
DFID	Department for International Development
DRM	Disaster Risk Management
DRR	Disaster Risk Reduction
ERC	Emergency Relief Cell
ERRA	Earthquake Reconstruction and Rehabilitation Authority
FAO	Food and Agriculture Organisation
FGD	Focus Group Discussion
FRC	Federal Relief Commissioner
ICRC	International Committee of the Red Cross
IFRC	International Federation of the Red Cross
ISDR	International Strategy for Disaster Reduction
km	Kilometre
NCMC	National Crisis Management Cell
NCOC	National Command and Operation Centre
NDMA	National Disaster Management Authority
NDMF	National Disaster Management Fund
NDRMF	National Disaster Risk Management Framework
NGO	Non-Governmental Organisation
NHN	National Humanitarian Network
PBUH	Peace be Upon Him
PDMA	Provincial Disaster Management Authority

PERRA	Provincial Earthquake Reconstruction and Rehabilitation Authority
PHF	Pakistan Humanitarian Forum
SERRA	State Earthquake Reconstruction and Rehabilitation Authority
UK	United Kingdom
UN	United Nations
UNDP	United Nations Development Programme
UNISDR	United Nations International Strategy for Disaster Reduction
USAID	United States Aid
WFP	World Food Programme

List of Figures

List of Tables

CHAPTER 1

Why a Book on the Role of the Mosque in Disasters?

1.1 Introduction

Disasters give rise to a situation where people from different parts of the world, quite unfamiliar with each other, come into contact to save lives, provide necessities such as food and shelter, rebuild homes and enable community recovery. Humanitarian and development organisations have concerns about keeping neutrality, staying safe, ensuring respect for local sensitivities and winning the necessary support of communities to conduct their job. At many places, during these challenging times, community-based religious institutions such as churches, mosques and temples are still but a practical choice for reaching people living nearby to fulfil their needs. Mostly, run by communities through charity in a decentralised manner, these institutions and organisations enjoy community trust and ownership.

Although according to the Pew Research Centre (2012), 84% of the world population affiliates to one of the many religious faiths, religion is far from being mainstreamed in the work of large development and humanitarian organisations where actors frequently refrain from work with religious or faith-inspired groups and organisations. In part, this is due to a lack of religious literacy in these organisations'. The policymakers both in public and private sectors, development practitioners and media personnel do not have a due understanding of religion. Instead of looking

A. R. Cheema, *The Role of Mosque in Building Resilient Communities*, Islam and Global Studies,
https://doi.org/10.1007/978-981-16-7600-0_1

at the religious interpretations employed by diverse groups for their vested interests, religion itself has been viewed as a driver of conflicts. However, due appreciation is lacking that religion stays intertwined with politics and development.

However, religion has recuperated considerable ground in the last half of the century after being relegated in stature. In both developed nations such as the USA and UK and many developing countries, religion continues to play transformational roles in the social, economic and political spheres of the lives of people. The current surge in academia about the role of religion in development was not initiated by academia but due to the practical presence of religiously inspired volunteer work noted by international institutions such as the World Bank and then by the governments in the aftermath of the 9/11 events.

Religion and community-based religious institutions influence the worldviews of their adherents in social, economic, political and environmental dimensions. Given the rise in extreme events and to reduce the extent of consequent losses, empirical and systematic evidence is needed to reduce the gap in knowledge about the vital role of community-based religious institutions. Still, we know little about the countless ways in which community-based religious institutions, groups and movements work in practice in a highly dynamic, adaptive and contextualised manner in different parts of the world. Disasters create a full range of opportunities for laying out a new landscape where new narratives can be built, old myths can be dismantled, and social realities can be manufactured. This knowledge is essential in building our understanding not only to plan, design and implement programmes for their safety given the rising extreme events but also to appreciate the full range of the influence of religion and religious institutions on different facets of the lives of people.

This is the gap this book attempts to bridge by presenting empirical evidence about the role of the mosque, a community-based religious institution, by documenting and analysing its role in rural settings in the aftermath of the 2005 earthquake in Pakistan. It provides detailed knowledge about the diversity of religious practices, less known otherwise, that occur in mosques and lead to varied responses in a post-disaster context.

This book makes a distinct contribution to the broader academic scholarship about the place of religion in development and its key role in the practice of disaster management. In particular, the book resonates closely with the actual and potential role of community-based religious institutions in the development and governance of the Muslim majority and minority countries. It makes a unique contribution to understanding the critical role that mosques play before, during and after a disastrous event and potentially other religious institutions, such as churches, temples and synagogues, can play in disaster management. To understand the complex relationships between communities and community-based religious institutions and leverage them to build safer communities, by presenting a compelling case of the role of the mosque, the book calls upon the involvement of the key actors from the state, civil society and private sector in disaster management.

This book provides essential reading for policymakers and emergency managers ranging from local to global ones. In terms of the global efforts for disaster risk reduction, the book argues for a collaborative approach with other actors following mutual understanding to empower the community-based religious institutions to realise their potential. Since outreach of the mosque message includes men, women and children, directly or indirectly, it may connect (coordinate, campaign, advocate and inform) and convince (organise, promote and encourage) and become an important channel for promoting disaster risk reduction activities and culture in Muslim communities. In an age of growing ethnic conflicts in different parts of the world, such an approach has an essential appeal to attract an audience as it may enable us to better prepare and survive future disasters both materially and spiritually.

Finally, the book adds to an ongoing conversation about the role of religion and religious institutions in widening development portfolios including that of peace, inequality, gender and environmental sustainability. As it happens, global economic, development and resilience agendas are no more exclusive domains of national, international and multilateral humanitarian and aid organisations. On the ground presence and the virtual power over human behaviour makes religion, religious institutions and their communities claim their due share in shaping development agendas that matter to communities. This book shows some of the pathways that can be leveraged in building resilient and inclusive societies by creating mutual trust and granting religion in the widening development agendas of the world.

1.2 Central Questions and Methods

This book explores the role of the mosque[1] in dealing with disasters in the aftermath of the 2005 earthquake in Pakistan. During this exploration, the book examines cultural, psychosocial, economic, social and political dimensions of the role of the mosque. To achieve this aim, the book addresses two primary questions:

1. What was the role of the mosque in relation to other key actors in the state, civil society and private sector during the response, relief, recovery, reconstruction and rehabilitation in the aftermath of the 2005 earthquake in Pakistan?
2. What are the potential future roles that the mosque and other similar community-based religious institutions can play in disaster preparedness, response, relief, recovery, reconstruction and rehabilitation that can be tapped in similar contexts in the future?

Using a case study approach based on qualitative methods of enquiry, three isolated and mountainous village communities having limited road access and located in the Mansehra district of the Khyber Pakhtunkhwa province of Pakistan were selected for an in-depth examination. In this case study design, different research methods including interviews, focus group discussions and participant observation were employed to collect primary data to develop an understanding of the role of the mosque in the lives of these communities.

The primary data collected during two fieldwork periods of five months in 2009 (including four weeks of participant observations) and 2010[2] by the author. In total, 83 interviews and nine focus group discussions were conducted. From villages and centres in the case study district, five mosques, seven imams, 13 women and 38 men were involved in interviews and focus group discussions in three villages named Banda-1,

[1] In this study, the expression "the mosque" is used to refer to the mosque as an institution, unless referring to a particular mosque building in a location.

[2] The first fieldwork period was April–July 2009. The second fieldwork period was mid-April to mid-May 2010. This section is based on my work during doctoral thesis titled "exploring the role of the mosque in dealing with disasters: A cases study of the 2005 earthquake in Pakistan" at Massey University, New Zealand available at http://hdl.handle.net/10179/4080.

Banda-2 and Banda-3 for anonymity. The remaining 40-interviews were conducted in six districts of Pakistan: Abbottabad, Hafizabad, Islamabad, Mardan, Muzaffarabad and Rawalpindi to achieve in-depth insight by visiting places and research participants within given resource and time constraints. Being aware of the complexity of disasters, a broad range of actors were recruited as research participants: from the government (central, provincial and district levels) including Pakistan army personnel; the private sector (local entrepreneurs, contractors and national and international consultancy firms); academia; research institutes; independent field experts; local and national non-governmental organisations (NGOs); and international multilateral organisations and journalists (from local and international media). Research participants were selected through purposive sampling. Data analysis was conducted through continuous review, coding and category development based on emerging themes. A triangulation method was used to validate and verify information received from primary and secondary sources.

The analysis in this book draws on this fieldwork conducted in 2009 and 2010, continuous observation and the latest secondary data from government documents. It situates the role of a mosque in a disaster context and draws its parallels to the ongoing role of religious institutions in fighting with COVID-19 pandemic in Pakistan and across the world. Combining the historical, decade-old primary data with the secondary data sources about the role of the mosque offers an interesting case study and also helps to frame the role of religion in the widening development agendas.

1.3 Disaster Management Cycle in Pakistan

Varying terminologies have been used for different phases of the disaster cycle in different countries such as New Zealand, Pakistan and the United States of America (USA). As a result, it's critical to figure out exactly what a phase in the disaster cycle implies in Pakistan. In the USA, the disaster cycle consists of four phases: mitigation, preparedness, response and recovery (Federal Emergency Management Association, 2010b). The 4Rs of a disaster cycle used in New Zealand comprise reduction, readiness, response and recovery (Ronan & Johnston, 2005). Although these terminologies appear similar, the terms differ in each country's context. For instance, in New Zealand, the term "recovery" is used holistically and the recovery phase is considered to include mitigation (Ronan & Johnston,

2005). In the USA context, recovery is more oriented to the rehabilitation of the damaged and destroyed infrastructure after a disaster. Likewise, the term "mitigation" differs in the USA and New Zealand. Keeping in mind the concept of the 4Rs, there is no mention of the term "mitigation" in the Civil Defence Emergency Management Act 2002 in New Zealand (Ministry of Civil Defence & Emergency Management, 2005) although it is referred to in terms of avoidance of risks in the "National Civil Defence Emergency Strategy" (Ministry of Civil Defence & Emergency Management & Affairs, 2007). On the other hand, there is a strong emphasis on mitigation in the Robert T. Stafford Disaster Relief and Emergency Assistance Act 1988 (Government of the United States, 1988). In the USA, the term "mitigation" is used in the sense of adopting measures relating to avoidance of future disasters preferably before a disaster (Federal Emergency Management Association, 2010a), which is different from the New Zealand interpretation of mitigation as patch up. In the USA mitigation means "reducing or eliminating long-term risks to people and property from hazards and their effects" (Federal Emergency Management Association, 2011).

In the context of Pakistan, Disaster Management Act 2010 specifies four phases of the disaster cycle: preparedness (1), response (2), recovery and rehabilitation (3) and reconstruction (4) (Disaster Management Act 2010, 2011). The National Disaster Risk Management[3] Framework (NDRMF) did not refer to the disaster cycle specified in the ordinance and later in the Disaster Management Act 2010 and used different terms including rescue and relief, response, recovery, mitigation, prevention and preparedness (National Disaster Management Authority, 2007). These terms have not been defined as such in the framework and the Act 2010. Table 1.1 shows the disaster cycle for Pakistan. The meanings of the terms—preparedness, response and recovery—as used in Table 1.1 have been adapted from the United Nations International Strategy for Disaster Reduction (UNISDR) (UNISDR, 2009) because the framework was prepared with the technical advice of the United Nations Development Programme (UNDP) of Pakistan. However, the UNISDR does not include definitions of rehabilitation and reconstruction. These terms explained in this book are based on their use by different people to whom I interacted during fieldwork, as well as on preliminary notes recorded in

[3] The term "disaster risk management" (DRM) is applied to depict "pro-active process of risk assessment and risk reduction" (UNISDR, 2009).

Table 1.1 Disaster cycle as per Disaster Management Act 2010 in Pakistan

Disaster phase	*Description*
1. Preparedness	The knowledge and capacities developed by governments, professional response and recovery organisations, communities and individuals to effectively predict, respond to and recover from impacts of current hazards
2. Response	The provision of emergency services and public aid during or at once after a disaster to save lives and keep public safety. It includes relief activities such as the provision of temporary basic life facilities such as food, shelter and medicine before recovery begins
3. Recovery and Rehabilitation	The temporary restoration of partially damaged public facilities, livelihoods and living conditions of the disaster-affected population to bring life back to normal
4. Reconstruction	The physical reconstruction of destroyed infrastructure, particularly government infrastructure like offices, hospitals, schools and bridges which may require a longer period based on the principle of "build back better" including measures for mitigation and reduction of disaster risk factors

Source Author

the Earthquake Reconstruction and Rehabilitation Authority's (ERRA) Community-Based Disaster Risk Management Programme documents (Earthquake Reconstruction & Rehabilitation Authority, 2009). ERRA was established as a statutory body on October 24, 2005, mainly to take up the enormous task of rebuilding the earthquake-affected region (Earthquake Reconstruction and Rehabilitation Authority, 2010).

This book, however, uses a modified disaster cycle as explained in Table 1.1. Although there are dedicated government institutions meant for relief, the relief phase has been merged into the response and is not mentioned as a distinct phase in the Act 2010 (Table 1.1). The disaster management cycle specified in the ordinance and later in the Act was influenced due to the developments in the wake of the 2005 earthquake such as the establishment of ERRA. Therefore, the classification of government disaster-related actors in this book is based on their functional responsibilities as explored from primary and secondary sources during the fieldwork. The four phases of the disaster cycle used in this

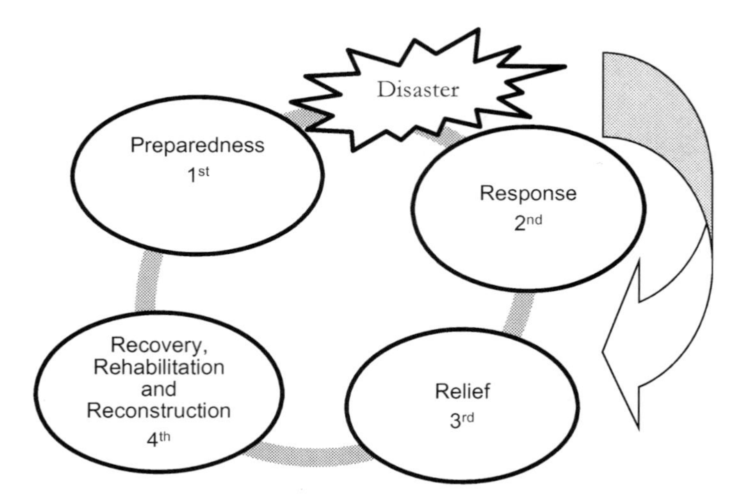

Fig. 1.1 Disaster management cycle in Pakistan's context as used in this book (*Source* Author)

book are preparedness (1st), response (2nd), relief (3rd) and recovery, rehabilitation and reconstruction (4th) as shown in Fig. 1.1.

It is important to note that disaster phases are not mutually exclusive because social settings are not homogeneous and different individuals and groups can be in different phases of a disaster at the same time (Neal, 1997). Conceptually, it is difficult to separate these phases as they are not neatly distinguishable from each other (McEntire, 2007). Disaster phases are therefore used as an organising concept to highlight the role of the mosque in this book and are not considered deterministically.[4]

[4] This section is based on my work during doctoral thesis titled "exploring the role of the mosque in dealing with disasters: A cases study of the 2005 earthquake in Pakistan" at Massey University, New Zealand available at http://hdl.handle.net/10179/4080.

1.4 Development, Natural Hazards, Pakistan and This Book

The 2030 Agenda, Sustainable Development Goals (SDGs) reaffirm the urgent need to reduce the risk of disasters and disaster losses as espoused through the Third UN Conference on Disaster Risk Reduction (UNISDR, 2015; United Nations, 2018). Specific opportunities to achieve SDGs through reducing disaster risk including by reducing exposure and vulnerability of the poor to disasters or building resilient infrastructure. Following several SDGs targets can contribute to reducing disaster risk and building resilience, even where disaster risk reduction is not explicit. Under SDG 4 related to promoting education for sustainable development by building and upgrading education facilities and ensuring healthy lives, as well as targets under SDG11 (cities) and under SDG 9 (building resilient infrastructure) reaffirm the interrelationship between disaster risk reduction and sustainable development.

The people of Pakistan, located in the Himalayan region, especially those in the northern part, are exposed to several natural hazards like earthquakes, floods, glacial lake outbursts, droughts, cyclones, storms, landslides and avalanches. The country is frequently affected by a range of disasters including recurrent flooding in the monsoon resulting in loss of lives, livelihoods and assets every year. Seismicity is especially high in the northern and western parts of the country.

Though disasters continue to offset development gains, the relationship between development, natural hazards and disasters are well understood and reflected in several government policy documents such as Disaster Risk Reduction Policy (2013), Climate Change Policy (2013) and National Disaster Management Plan (2013–2022) (National Disaster Management Authority, 2019). Similarly, the National Disaster Response Plan (2019) shows explicit linkages and aligned with global frameworks including Sendai Framework (2015–2030), Sustainable Development Goals (2015–2030) and Paris Agreement 2015.

Immediate pressures that hinder investment by the government in disaster risk preparedness and risk reduction measures include rapid population growth, unplanned urbanisation and industrialisation, poverty, climate change impacts, and large reliance on agriculture and livestock for livelihood and food (National Disaster Management Authority, 2019). The state capacity remains weakest in terms of human capital and financial resources, particularly at the district level and that too in poverty-stricken remote areas of the country (Rahman et al., 2015).

In this context, this study takes a closer look at the lives of the poor communities of Mansehra (district of the Khyber Pakhtunkhwa province of Pakistan) who labour to allocate resources for the reduction of future risks at the cost of their day-to-day struggle to meet immediate needs. It explores how these communities build on their religious beliefs and practices through the institution of the mosque in their disaster-prone environment with minimal expectation for any external help. Against this background, this research examines the significance of the mosque, a local community-based institution, in the everyday lives of the poor. It also studies the social and physical environment that may influence local communities to combine their energies to respond to and face everyday vulnerabilities on a self-help basis for their survival. Overall, this book seeks to highlight the relationships between disasters, development and community-based religious institutions in influencing the well-being of the local people, both in response to hazard events and to protecting people's lives and assets from future hazard events.

1.5 The Organisation of the Book

Chapter 2 addresses the critical challenges to engagement with religion and religious institutions such as neutrality, transparency and the prompted need for a secularised humanitarian development agenda versus potential and actual advantages of cultural proximity are confronted in this chapter. The debate juxtaposes the opportunities and threats to such sensitive engagements and lends the view that the opportunities are far greater than threats. It also unpacks the political controversy about the use of mosques as a harbouring place for terrorism. Lastly, the chapter brings together key examples of the role of the mosques concerning social dimensions of life such as water conservation and fighting poverty and blindness and potential roles during disasters.

Chapter 3, briefly, introduces the disaster management institutions and structures concerning the role of the government, private sector and civil society that existed at the time of the 2005 earthquake. This background is helpful to illuminate the actual role of the mosque in the aftermath of the 2005 earthquake in Pakistan, discussed in the next chapter.

The actual role of the mosque during the response, relief, recovery, reconstruction and rehabilitation is discussed through the primary data collected from affected people, Imams, the staff of the local, provincial, national and international organisations who worked in different phases of the recovery operation in this chapter. The chapter also brings forth the potential roles of the mosque as identified by the research participants and corresponded to the postulated role for a community-based institution by the United Nations International Strategy for Disaster Reduction.

The last chapter synthesises the role of religion in development and disaster studies discourse about the actual and potential role of the mosque and other religious institutions in the lives of people. Enriching the scholarship, it distils theoretical and practical contributions of the mosque as a physical, spiritual and social place and other community-based religious institutions for improving the knowledge and practice of disaster management. It also illuminates the widening role of religion in development avenues through this case study of the mosque. This is a new example that illuminates place-based religious practices that influence, transform and negotiate the practice of disaster management of the nearby communities in both Muslim majority and minority countries. Besides, this also reinforces the case for broader engagement with all community-based religious institutions such as churches, synagogues and temples.

References

Earthquake Reconstruction and Rehabilitation Authority. (2009). *Information booklet for Union Council Disaster Management Committee (UCDMC)*. Earthquake Reconstruction and Rehabilitation Authority (ERRA), Government of Pakistan.

Earthquake Reconstruction and Rehabilitation Authority. (2010). *Earthquake Reconstruction and Rehabilitation Authority (ERRA) - About us*. http://www.erra.pk/aboutus/erra.asp#HB

Federal Emergency Management Association. (2010a). *Mitigation*. Federal Emergency Management Association (FEMA).

Federal Emergency Management Association. (2010b). *Mitigation best practices portfolio*. http://www.fema.gov/plan/prevent/bestpractices/index.shtm

Federal Emergency Management Association. (2011). *FEMA mitigation and insurance plan*. Federal Emergency Management Association (FEMA).

Disaster Management Act 2010. (2011). http://web.ndma.gov.pk/files/NDMA-Act.pdf

Government of the United States. (1988). *Robert T. Stafford disaster relief and emergency assistance Act (Public Law 93–288) as amended upto June 2007* (F. E. M. A. (FEMA), Ed.). http://www.fema.gov/about/stafact.shtm

McEntire, D. A. (2007). *Disaster response and recovery: Strategies and tactics for resilience*. Wiley.

Ministry of Civil Defence & Emergency Management. (2005). *Focus on recovery: A holistic framework for recovery in New Zealand* (M. of C. D. & E. Management, Ed.). Ministry of Civil Defence & Emergency Management (MCDEM), Government of New Zealand. http://www.civildefence.govt.nz

Ministry of Civil Defence & Emergency Management, & Affairs, D. of I. (2007). *National civil defence emergency management strategy*. Ministry of Civil Defence & Emergency Management (MCDEM), Government of New Zealand.

National Disaster Management Authority. (2007). *National disaster management framework of Pakistan*. National Disaster Management Authority (NDMA) Government of Pakistan.

National Disaster Management Authority. (2019). *National disaster response plan* (Issue 1). http://cms.ndma.gov.pk/

Neal, D. M. (1997). Reconsidering the phases of disaster. *International Journal of Mass Emergencies and Disasters, 15*(2), 239–264.

Pew Research Centre. (2012). *The global religious landscape*.

Rahman, A., Shaw, R., & Khan, A. N. (2015). *Disaster risk reduction approaches in Pakistan*. Springer.

Ronan, K. R., & Johnston, D. M. (2005). *Promoting community resilience in disasters: The role for schools, youth, and families*. Springer.

UNISDR. (2009). *UNISDR terminology on disaster risk reduction*. United Nations International Strategy for Disaster Reduction (UNISDR). http://www.unisdr.org/eng/library/UNISDR-terminology-2009-eng.pdf

UNISDR. (2015). *Sendai framework for disaster risk reduction*. UNISDR. https://www.preventionweb.net/files/43291_sendaiframeworkfordrren.pdf

United Nations. (2018). *2018 Review of SDGs implementation: SDG 11 – Make cities and human settlements inclusive, safe, resilient and sustainable*. https://sustainabledevelopment.un.org/content/documents/197282018_background_notes_SDG_11_v3.pdf

CHAPTER 2

Disasters and Religious Institutions

2.1 Introduction

Disasters give rise to a situation where people from different parts of the world, quite unfamiliar with each other, come into contact to save lives, provide necessities such as food and shelter, rebuild homes and enable community recovery. Humanitarian and development organisations have concerns about how to keep neutrality, stay safe, ensure respect for local sensitivities and win the necessary support of communities to carry out their job. In many places, during these hard times, community-based religious institutions such as churches, mosques and temples are still but a practical choice for reaching people living nearby to fulfil their needs (Feener & Fountain, 2018; Ngin et al., 2020). Mostly, run by communities through charity in a decentralised manner, these institutions and organisations enjoy community trust and ownership (A. R. Cheema et al., 2014; Gianisa & Le De, 2018). This chapter presents an overview of religious views and approaches to disasters and why religious institutions remain an important stakeholder during and before disasters. Also, it covers some of the key challenges that hinder an effective engagement between religious institutions and large international and national humanitarian organisations.

A. R. Cheema, *The Role of Mosque in Building Resilient Communities*, Islam and Global Studies,
https://doi.org/10.1007/978-981-16-7600-0_2

2.2 Religious Views and Approaches

People from different religious, supernatural and superstitious contexts have often labelled disasters as "acts of God" in the past. Many thought that disasters were to satisfy superstitious designs, and therefore, they attributed the loss of life and property to acts of gods and deities. Even the origin of the word disaster has superstitious undertones. The original Latin meaning is a star. In this context, disasters have been related to the movement of stars, in particular, "ill-starred" (Gibson, 2006). People have thus often ascribed their misfortunes to heavenly forces and considered calamities as a punishment from gods.[1]

This does not, however, mean that such societies have been altogether fatalistic. Almost all religions suggest means such as prayers, sacrifices and charities to their followers to please their deities to ward off misfortune, calamities and disasters (Stern, 2007). Sometimes, religions provide examples and illustrations to their adherents for disaster prevention and preparedness measures. The story of the Prophet Yusuf (PBUH)[2] in the Holy Quran (Chapter 12), referred to as Prophet Joseph in Christian and Jewish scriptures, describes food rationing and storage to save people from drought and famine, around 1500 B.C. (Arutz Sheva, 2009). Briefly, as per the story in the Holy Quran (Chapter 12), the King of Egypt dreamt that seven fat cows were devoured by seven lean cows and seven green ears of corn were replaced by seven dry ears of corn. Prophet Yusuf (Joseph) interpreted the dream as seven years of a good harvest to be followed by seven years of severe drought. The King was impressed with the Prophet's interpretation of the dream and appointed him to look after the warehouses of Egypt. The dream came true and enough food was stored in the first seven years of abundant crops. This famine planning saved the people of Egypt during the seven years of severe drought that followed. This example also shows that planning for disasters has been practised by human beings for centuries and is not something that evolved in the twentieth century (Quarantelli, 2009).

[1] This section is based on my work during doctoral thesis titled "exploring the role of the mosque in dealing with disasters: A cases study of the 2005 earthquake in Pakistan" at Massey University, New Zealand available at http://hdl.handle.net/10179/4080.

[2] A salutation spoken and written with names of all Prophets in Islamic tradition, abbreviated as PBUH (peace be upon him).

Even in ancient history, in some instances, people showed an organised response to impending disasters. Coppola (2007) notes the example of two towns in Italy, Herculaneum and Pompeii, which faced volcanic eruption in AD 79 from Mount Vesuvius. The leaders of Pompeii coordinated the evacuation of the residents of the town many hours before pyroclastic flows approached their town and thus saved many lives, even though the city was buried in the ash. The inhabitants of Herculaneum, however, could not survive since the town was at the base of the Mount and pyroclastic flows overtook the town in no time.

Religious interpretations have shown certain alterations as a result of learning from experience and as knowledge has grown over time. For example, an English clergyman, Thomas Robert Malthus (1766–1834), shared his view of disasters as acts of God while suggesting that nature would take away the excess population if human beings did not observe moral restraint. He argued that nature would provide for as many people as it could feed. If human beings exceed the availability of food on earth, nature would balance the proportion of human beings and food by removing the surplus number of humans through floods, droughts, diseases, famines and wars (Malthus, 1958).

In the eighteenth century, a major earthquake struck the Portuguese city of Lisbon on November 1, 1775, resulting in widespread destruction (Quarantelli, 2009). The elite interpreted the Lisbon earthquake on scientific lines, whereas the general population thought it to be an act of God due to people's sins[3] (Grandjean et al., 2008). The famous influential French philosopher, Voltaire (1694–1778) quoted in Grandjean et al. (2008), authored a poem challenging the perception that Lisbon was destroyed because of sins:

> Did fallen Lisbon deeper drink of vice.
> Than London, Paris, or sunlit Madrid?
> In these, men dance; at Lisbon yawns the abyss.

Voltaire compared Lisbon with London, Paris and Madrid asking the question of whether the former was worse in immorality than the latter. The Lisbon earthquake had a profound impact. Perhaps for the first time in the world, the government introduced earthquake-building codes

[3] Here it refers to a human act that is wrong according to the religious laws of any particular faith.

during the rebuilding of the city (David K. Chester & Chester, 2010). This may have been the beginning of scientific interpretation, in contrast to an "acts of God" notion of disasters. It seemed to be the beginning of a new dimension in human thought that there was a need to have some kind of control over nature and not be satisfied with the idea of divine retribution. "The idea of risk management emerges only when people believe that they are to some degree free agents", argues Bernstein (Bernstein, 1996). While scientific explanations regarding the Lisbon earthquake were suggested, the influence of religious interpretations of disasters to varying degrees persisted at different places, and the scientific interpretation of disasters was not widespread and common knowledge.

Even more recently, religious interpretations have still prevailed and many people, both from the developing and developed world, continue to attribute disasters to God. For example, many Indonesian people interpreted the 2004 tsunami as a test of their love for God or a punishment for their sins (Brummitt, 2006). Similarly, in the aftermath of Hurricane Katrina in the USA, some called it divine wrath because of immoral acts (particularly same-sex marriages) and some saw it as divine judgement on atrocious human behaviour (Vaught, 2009).

Religious interpretations of disasters were, however, most common before the mid-twentieth century when societies dealt with disasters on an ad hoc basis and there was no tradition of the systematic study of disasters (Quarantelli et al., 2007). A multi-hazard approach considering comprehensive measures to safeguard societies from all kinds of hazards was absent. There are, however, historical examples of planned efforts to reduce disaster risk. The 1667 Act for Rebuilding London adopted in the aftermath of the great fire of London in 1666 is an example of a planned approach to reduce the impact of fire hazards in the future (Platt, 1998).

Overall, for many people around the globe, religious views and practices still constitute an important explanation for disasters, but this perspective has not had much focussed attention of the scholars until the early 2000s. In the context of Pakistan, disasters have generally been perceived as a test or a punishment from God on society, and community-based religious institutions such as the mosque have played an important role, as explored in detail in Chapter 4.

2.3 Religion Survived the Onslaught of Various Ideological and Political Challenges

Community-based religious networks remain among the first line of responders at the time of disasters, particularly in remote and far off mountainous places (A. R. Cheema et al., 2014). Majority of the respondents in a study by Suri (2018), Buddhist and Muslims in Ladakh and Kashmir run to monasteries and mosques where they feel safe. A decade ago, in a special issue of the journal "Religion" on the themes of religion, natural hazards and disasters, Gaillard and Texier (Gaillard & Texier, 2010) pinpointed that disaster studies neglected the role of religious institutions for a long time. This was even though religious institutions have been serving the people long before the European tradition of humanitarian aid, as stated in the Oxfam (Cairns, 2012) paper:

> The humanitarian project is not just a European tradition. It is rooted in the universal behaviour to help other human beings in distress. It has been encapsulated in all faiths, from Dana, one of Hinduism's and Buddhism's vital practices, to Islam's Zakat, and Christian charity. It is no coincidence that local religious organisations are at the forefront of providing relief.

Religious institutions have been contributing to different phases of disasters including response, recovery and rehabilitation at the local level, where religion has significant influence in shaping perceptions of vulnerable communities (D. K. Chester et al., 2008). During times of disasters, religious institutions contribute to the disaster mitigation drive in several ways such as feeding hungry victims, providing shelter and supporting the communities holistically, along with other stakeholders. Religious institutions have played an important role in developing social cohesion by building social and safety networks within communities (Bano & Nair, 2007; Candland, 2000). Wisner (2010) asserts that religious communities, groups, institutions and leaders have an "untapped potential" for the task of disaster risk reduction at the local level. He emphasises that religious groups and organisations are usually the first responders because of their immediate availability and strong local networking. However, he finds the need for the engagement of religious communities in community preparedness for disaster prevention so that the untapped potential of these communities could be used for saving lives and reducing vulnerabilities and economic setbacks.

Religion and community-based religious institutions influence the worldview of their adherents in social, cultural, economic, political and environmental dimensions (Holmgaard). Local or indigenous knowledge takes a holistic view of life and environment and thus engages with several qualitative contextual variables such as kinship, mutual respect, sharing and reciprocity in the world and hereafter (particularly in faith-based societies). Berkes (2008) argues that indigenous knowledge conceives holism by considering many factors qualitatively and perceives a unity in life and nature while western sciences tend to concentrate on a small number of variables quantitatively. Bankoff (2004) reinforces that nature and disasters are incorporated in everyday life and thus there appears a social construction of response to hazard and vulnerability.

Religious institutions are far from being worship places only as they play real and potential roles in shaping and aiding a community's preparedness, response and recovery from a disaster event. However, engagement with religious institutions is not a straightforward matter of choice. A nuanced understanding of the under-studied political, economic and social contexts is essential. This knowledge is essential in building our understanding not only to plan, design and implement programmes for their safety given the rising extreme events, but also to appreciate the full range of the influence of religion and religious institutions on different facets of the lives of people (Sheikhi et al., 2020).

Religion has recuperated considerable ground in the last half-century after being relegated in stature. In both developed nations such as the USA and UK, as well as many developing countries, religion continues to play transformational roles in the social, economic and political spheres of the lives of people (Feener & Fountain, 2018). The current surge in academia about the role of religion in development was not initiated by academia but due to the practical presence of religiously inspired volunteer work noted by international institutions such as the World Bank and then by the governments in the aftermath of the 9/11 events.

Beyond disasters, given the devastation produced by large-scale construction projects, there are calls for greater awareness of the importance of religious and spiritual places as vital components of human well-being. Broadly categorised as cultural heritage, Griffiths et al. (2020) emphasise understanding the importance of nature-based cultural values including religious and sacred places to people's well-being while implementing large development projects. They also conclude that these values differ between people and groups of people. However, comprehensive engagement based on the "no worse off" principle is vital to ensure fair and socially inclusive development. This requires in-depth knowledge and trust of the local people and different variations within different communities. This knowledge and trust cannot be gained with the occasional and transactional nature of relationship and interaction that prevails among most of the government and other disaster-related organisations.

A study on the role of three major religions in South Korea Ha (2015) shows how Christianity, Buddhism and Confucianism have been serving Korean communities during disaster response and relief, mostly care-oriented management. For improving disaster preparedness and mitigating human losses, economic damages and psychological impacts, both, the Government of South Korea and the leadership of these religions need to work together closely. The central government agencies working on disaster management might consider including related local religious narratives in disaster education and training materials. Others such as Sheikhi et al. (2020) also call for mitigation-oriented management for religious institutions to increase the effectiveness of disaster response and recovery.

Advances are also taking place in examining the nuanced role of religious actors in the field of psychology in disaster contexts. Hirono and Blake (2017) examine the role of American (Christian in this case) and Japanese (Buddhist in this case) clergies' perception of their role in reducing post-traumatic stress disorders (PTSD) in the aftermath of natural hazard triggered disasters. While both clergies inculcate hope among survivors and help them to cope with PSTD, they find differences in their modus operandi. American clergies pay attention to reducing pain and supplying comfort while Japanese clergies concentrate on listening to the disaster surviving families and praying for the lost ones. To better address the psychological needs of the disaster survivors, the authors call for close working within different clergies, between clergies and mental health professionals.

Although the integrative approach to disasters has attempted to consider social, cultural and economic factors since the 1980s, the role of religious institutions along with the role of religion itself remains overshadowed and underestimated in the disaster studies literature (Candland, 2000; David K. Chester, 2005). Even now, much below their full potential, most of the government and international disaster-related organisations only involve with religious and faith-based organisations to the extent of disaster response and relief, disregarding engagement for long-term disaster recovery and preparedness (Gingerich et al., 2017; Sheikhi et al., 2020). Overall, there is a gradually increasing realisation among both religious and secular development organisations to explore ways to engage with each other for achieving better results of their common efforts for helping humanity. While progress has been made about the recognition of religious institutions, methodologies and complexities that need to be understood for this recognition to translate into effective engagement have yet to happen (S. Deneulin & Zampini Davies, 2017; Séverine Deneulin & Rakodi, 2011).

2.4 Challenges to Engagement with Religious Institutions

This section addresses the critical challenges to engaging with religion and religious institutions such as neutrality, transparency and the prompted need for a secularised humanitarian development agenda versus potential and actual advantages of cultural proximity. The debate in this section would juxtapose the opportunities and threats to such sensitive engagements.[4]

2.4.1 Similarities Between Goals of Religious and Secular Organisations

The search for ideological similarities, such as those driven by the commonality of goals, between the conventionally opposing entities—religious versus secular organisations—is another reason for closing gaps and working together at national and international levels. In a broader sense, the ideology informing both secular and religious development is to alleviate human suffering.

Multilateral organisations such as DFID began to forge closer links with Muslim groups. Building on Muslims' religious principle that "humanity is like a body; discomfort or pain in any part of the body causes the whole body to suffer" (Department for International Development, 2001).

[4] Some parts of this section draw on my work during doctoral thesis titled "exploring the role of the mosque in dealing with disasters: A cases study of the 2005 earthquake in Pakistan" at Massey University, New Zealand available at http://hdl.handle.net/10179/4080.

UNICEF cherishes its long-term partnership with religious communities and faith organisations on a wide range of issues that affect children (UNICEF, 2017). It provides a framework for effective engagement while acknowledging the wide diversity in religious communities. The framework calls for understanding values, structures and leadership as the first step. Later, it focusses on working with shared values and a right-based approach.

Despite these clear acknowledgements and efforts to bridge some of the gaps between religious and secular humanitarian organisations over time, the 2004 tsunami showed tensions on the ground. McGregor (2010) examined roles and secular policies (not to engage with religious institutions such as mosques in Aceh) of the international aid organisations of New Zealand (New Zealand Aid) and Australia (Australian Aid) in Aceh after the 2004 tsunami. He concluded international aid and humanitarian organisations almost missed huge opportunities for supporting physical and psychosocial recovery and increase the marginalisation of the affected communities by refusing to acknowledge religious authority or engage with religious institutions.

After and during the COVID-19 pandemic, this collaboration and joint working further progressed with the release of new guidance advising faith leaders and communities how to practice faith safely, fight misinformation and support children and vulnerable population (Wilkinson et al., 2020). Faith and Positive Change for Children (FPCC) initiative was conceived in 2018 to move to a genuine multi-sectoral approach from a single sector, siloed, ad hoc and sometimes instrumentalist approaches of faith engagement in development work. Joint Learning Initiative on Local Faith Communities (JLI) is a knowledge partner in this initiative. Both the worlds' largest interfaith network, Religions for Peace (RfP) and UNICEF officially launched FPCC in 2019.

2.4.2 *Sensitivity*

Religion is usually a sensitive matter in societies. It is becoming difficult to dissuade suspicion attached to the development agenda of international development agencies working in recipient countries (Bonney & Hussain, 2001). There have been instances of attacks on the headquarters of the central disaster authority also occurred in the post-tsunami Ache due to ignorance of local religious sentiments (including NGOs avoiding to build sacred places as mosques and due to the presence of proselytising groups)

(McGregor, 2010). A USA faith organisation, World Help, announced that it was to move 300 Muslim orphan children from Aceh to a Christian children's home in Jakarta with a plan to inculcate the gospel and faith in Christ in their minds. This did not happen since the community became aware of it and was successful in resisting the move.

Overall, religion and religious norms are often sensitive issue to communities. Any intentional or unintentional confrontation with religious institutions can spark conflict among stakeholders and may erode the necessary confidence of the local community and other development partners. The involvement of religious institutions such as churches, mosques, synagogues and temples through their faith leaders could be instrumental in resolving issues of sensitivity and thus improving the relationship among different stakeholders involved in disaster management.

The platform of religious institutions has been successfully employed for addressing broader issues of cultural awareness about ethnic minorities that can help when responding to disaster situations. The Western Australia Fire and Emergency Services Authority conducted a workshop in coordination with Muslims and mosques to sensitise its staff to different matters relating to Muslims, 1.7% of the Australian population, such as entering mosques and how to treat Muslim women when physical contact might be necessary (Fozdar & Roberts, 2010).

2.4.3 Proselytising, Discrimination and Neutrality

Concerns about proselytising, prejudice and neutrality have contributed to other players' cautious stance in their interactions with religious institutions.

Twigg (2004) cautions about the possibility of discrimination in the distribution of relief aid based on religion and faith. In a case study of the village Katni in Bangladesh, Hartman and Boyce 1983 relate the higher death toll of the minority population (Hindus) in a Muslim majority community during times of food shortages (due to drought or bad harvest). Similar instances of discrimination occurred with the Hindu (minority groups) affected by the 2010 flooding in Pakistan (Suhail, 2010). These affected Hindus complained about being provided with cow meat, which Hindus do not eat, and Muslim's left-over food. Later, Hindu temples in big cities like Karachi accommodated most of the affected Hindus and provided them with food, shelter and medicine.

To mention an example of Edhi Foundation, a large social welfare organisation is a success story in Pakistan. It was founded by Late Abdul Sattar Edhi from Karachi, a Muslim, and has a reputation of serving all segments of the society without discrimination.

2.4.4 Case Study of Samoa

Holmgaard (2019) shows recent evidence of how pastors and missionaries belonging to the new churches used post-disaster (tsunami) for converting people to Christianity during disaster recovery in Samoa. Here, people actively looked to interpret tsunami through religious teachings. Despite individual differences, two distinct themes, first, members of the mainline churches considered tsunami as a punishment for their sins, and second, members of the new churches attributed the tsunami to the Second Coming of the Christ, emerged. It was this group, evangelical, harvested the time of stress and uncertainty for the people to convert them before the Second Coming of Christ. Proselytising was easier as these churches offered emotional and material support with compassion. Unlike those who converted to the new churches before the tsunami, most of those who converted after the tsunami belonged to low-income groups. In Samoa, Christianity is a dominant religion, and churches and pastors are deeply intertwined in social, political and economic spheres at national and village levels.

Holmgaard (2019) also demonstrates that the process of the conversion to new churches did not start after the tsunami. Even before the tsunami, the pastors and missionaries of the new churches were openly critical of the demands of the mainline churches, such as donations and ceremonial obligations. They were inviting Samoans to an individualistic style of relationship with God with minimal financial, kinship and ceremonial obligations to the church. They used the post-tsunami situation once people suffered from financial and mental stress to attract them to the new churches with a spirit of getting them ready before the Second Coming of Christ.

Many religions also stress the importance of believers being involved in charitable acts (Stern, 2007). In an incident of the stampede at Jodhpur Chamunda Devi temple in India, which claimed 186 lives, Sain (2008) reported that Muslims were the first responders and rescuers, ferrying victims to hospitals and queuing up to donate blood. When one of the Muslims was asked why he was helping Hindus, he replied "we are all creations of Allah Who does not differentiate between Hindus and Muslims". A similar instance of hospitality towards Hindus from Muslims inspired by religious teachings occurred during the 2011 floods in the Sindh province (Khan, 2011). At that time, the local government officials left the people on their own after the flood hit the town. A local mosque played the role of a community-based institution and the imam provided the much-needed leadership to benefit all local people, irrespective of the religious divide.

2.4.5 Disaster Risk Perception

Modernity and enlightenment in the post-industrialisation era have altered the general idea of ascribing disasters to God, spirits or religion. There is a broad academic consensus in the disaster studies literature that disaster risk perception is fundamental to the determination of people's responsiveness, preparedness and ultimately, to the growth of disaster prevention culture (McGeehan & Baker, 2017). The influence of religious institutions on disaster perceptions may be seen from two angles—a general view of disasters and belief in fate.

2.4.6 Different Views of Disasters

Despite increasing recognition of the views of the primary stakeholders in disaster management, the challenge is still how to moderate between modern and traditional knowledge. Traditional knowledge is usually based on fate, religion, belief, culture and history, while modern knowledge has its mainstay on scientifically verifiable evidence. To further add to this complexity, there is a large variety of views about disasters within a religious tradition. Chester et al. (2008) observed the case of Catholics in Southern Italy. Among Catholics, one school of thought has similar views on disasters to that of mainstream disaster studies. This school of thought, "liberationist", notes the disproportionate impact on the poor and marginalised in disasters and considers that disasters are the result of institutional sins rather than that of individuals. The other group of Catholics, Irenaean, still deems that natural hazards serve a larger good such as earthquakes helping mountains to grow and volcanic eruptions helping to establish planetary atmosphere.

Despite these divergent views, there was a broad sense of cooperation with authorities from the public and no confrontation was observed from religious institutions towards evacuation calls following various volcanic emergencies. Having these different kinds of interpretations of volcanic eruptions even within followers of the same religion, should not deter authorities from engaging with them.

Within these varying religious views of disasters in Christian traditions, there are strong opportunities for cooperation, integration and partnership between religious institutions and secular organisations for the reduction of disaster losses. The other aspect through which religious institutions influence public perception of disasters is the concept of fate. Like perception, fate also shows varying characteristic within and among different religious groups.

Paradise (2005) finds that community risk perception can critically affect the success of national or regional disaster policy and practice in faith-based societies. The author surveyed the city of Agadir, Morocco, which is comprised of Muslims. The author's focus was to assess the risk perception of the community 40 years after a deadly earthquake in 1960. The study found that out of 243 interviews, 56% of women and 51% of men did not answer the question about whether they expect another earthquake in their lifetime, instead of replying "Allahu a'lam" or "God is the wisest". This answer was given even though Agadir is situated on active faults. These respondents thought that it was *haram*[5] or prohibited to predict and speculate about earthquakes. They felt earthquake prediction is equivalent to fortune-telling. Such perceptions among the majority of community members are more than mere academic findings. Paradise (2005) cautions that the situation calls for a serious reflection on the success of disaster management policies.

Religious institutions and beliefs do not necessarily suggest communities become fatalistic and they are not enemies of modern knowledge per se as assumed by many secular humanitarian organisations (Gingerich et al., 2017). Several socio-economic factors affect the attitude and behaviour of communities towards their immediate environment and such tendencies are often misinterpreted by outsiders.

People having religious inclination can follow a "parallel practice" which means invoking divine help from their belief in God and adapting standard precautionary measures like secular communities do in times of disaster (D. K. Chester et al., 2008). Thus, about 7,000 Sicilians gathered

[5] Haram means forbidden or prohibited by Islam.

for a mass prayer under the leadership of the Archbishop of Catania, at a sanctuary in Belpasso to stop the lava flow from Mount Etna (Kennedy, 2001). On some occasions, people in southern Italy were even found to be keeping deities before the lava as a means to stop it, while also listening to and following the evacuation calls by the government authorities (D. K. Chester et al., 2008). Long-term and sustained engagement with the religious and faith communities can show different shades of interaction about what is apparent and what is not apparent. The river-side communities of Bangladesh ascribed their fate to Allah but did not turn fatalistic at all (Schmuck, 2000). Schmuck observed that the people living along the banks and islands of the Jamuna river followed a variety of coping strategies despite believing in God as the ultimate controller of their future. Using a participant observation technique rather than using a structured questionnaire to elicit the responses of people, the author observed that people moved to other forms of livelihood, used their saved stock of food from the winter crop and banked on extended family networks at times of flooding. The author's finding is in contrast to the earlier view that believing in God necessarily meant one was a fatalist. Later, Hutton and Haque (2003) affirmed the findings of Schmuck (2000). They conducted a study of communities living along riverbanks in riverine zones of Bangladesh. The authors observed that most of the people stated that their future rested on the will of Allah but still they adopted all types of safety measures. Similar religious beliefs that did not turn communities into fatalists were observed while interviewing people in Cairo about their risk perception of an earthquake after the 1992 earthquake in Cairo. Homan (2003) found that religious explanation was meaningful to people and very few felt that this precluded practical action to take earthquake loss mitigation steps.

Overall, religious institutions are important partners in disaster management, since many people will continue to see disasters and disaster losses in religious terms. There are and will be opportunities for making wise choices and preparing communities for impending disaster at the time of religious-driven activities such as prayer gatherings through local faith leaders (Anthony Oliver-Smith, 2004; Gingerich et al., 2017).

2.4.7 *Opportunities for Psychosocial Support, Spiritual Healing, Resilience and Charity*

Many people tend to get close to religion at a time of stress such as disasters and calamities (Gibson, 2006; Guarnacci, 2016; Joakim & White, 2015). Not only that, attachment to religion offers a unique opportunity for fighting against stress, trauma and tension. A religious connection forms an important constituent in the framework of generalised resistance resources proposed by Antonovsky (1979). This framework refers to "any characteristic of the person, the group, or the environment that can facilitate effective tension risk reduction". "Culture is a distributed system of models and the cultural dimensions of mind are simply those mental models that derive from shared institutions", observes Shore (2002). Religious institutions affect and influence culture, as well as receive influence to varying degrees in each set of socio-cultural dimensions. Over a period, religious institutions become shared institutions that have a bearing on people's behaviour and their orientation towards life and its processes.

Resilience to shocks and disasters including COVID-19 encompasses the ability of societies to cope with, recover from and adapt to the effects and demands of uncertain present and future events in economic, social and environmental aspects (Leach et al., 2020).

Resilience is not a static attribute, rather it is shaped by culture (to which religion is a key element at many places), social networks and personal attributes (Cottrell, 2006). Communities with high vulnerability may still possess and reflect higher levels of resilience at the time of a disaster. The conventional belief that higher levels of vulnerability are associated with lower levels of resilience may not hold for all cases. Resilience is a dynamic attribute, which varies from place to place, culture to culture and people to people. A uniform degree of resilience should not be conceptualised, as resilience differs among individuals and communities (Pooley et al., 2006).

Religion has been a major source in creating resilience in individuals and communities in different kinds of post-disaster situation trauma and stress. For example, after Hurricane Katrina, attachment to religion and religious institutions by contemplating and visiting churches was a great source of strength and resilience for African-Americans coping with mental stress (Laditka et al., 2010). Similarly, another study exploring stress coping strategies among unaccompanied minors of asylum seekers in Ireland found that religion and going to church was the key source of creating resilience (Ni Raghallaigh & Gilligan, 2010).

Religious institutions motivate individuals to volunteer and help disaster victims and thus nurture a culture of altruism. In addition, religious institutions inspire individuals to be selfless and foster linkages and bridge social gaps in a society that may be diverse and at odds at times. Religious institutions offer support through a religious worldview, provide psychosocial healing, create resilience through a shared meaning of life and foster the process of normalisation of human behaviour. People of all ages and both sexes in different cultural and social settings have referred to religion as the main source of resilience for them in various stressful conditions.

The discussion so far has shown some of the complexities concerning how these roles of religious institutions unfold in different contexts. After having this overview of the role of religious institutions, in a disaster context, the next section introduces one such religious institution, which is the case study for this book, the mosque.

2.5 Introducing the Mosque

It is important to explain how the term "mosque" is used in this book because it is at the core of this study. The succeeding discussion provides a brief overview of the role of the mosque and its political implications, an explanation on the classification of the mosque as an organisation or an institution, a faith-based organisation or a religious institution, and finally, what type of institution or organisation it is.[6]

Primarily a place of worship, the mosque has been a spiritual, educational, cultural, social and administrative institution in Muslim societies, historically well-documented and recorded since the time of the Prophet Muhammad (PBUH) fourteen centuries ago (Mahr, 2005). Mosques have been a focal point for different activities such as the teaching of religious and general knowledge to men, women and children, housing and feeding of the poor, and conducting social gatherings like weddings. In

[6] Some parts of this section draw on my work during doctoral thesis titled "exploring the role of the mosque in dealing with disasters: A cases study of the 2005 earthquake in Pakistan" at Massey University, New Zealand available at http://hdl.handle.net/10179/4080.

addition, the mosque has been a place for war-injured soldiers and travellers to rest, negotiation with foreign delegates, and announcements of important decisions and sporting competitions. The imam of the mosque was usually the administrator of the area. However, the nature of the role of the mosque has changed over time, so now the mosque is more active in Muslim-minority countries than in Muslim-majority countries (Abdel-Hady, 2010). Generally, in Muslim minority countries, such as the majority of western countries, the mosque also provides a social and cultural space for Muslim communities, whereas these activities have shifted outside of the mosque in Muslim-majority countries.

Muslims can offer prayer at any clean place although Allah and His Prophet Muhammad (PBUH) have enjoined believers to establish prayers with a congregation in the mosque. It is an act of great merit to establish mosques. A verse of the Holy Quran (9:18) reads, "only those shall visit and maintain the mosques of Allah who believe in Allah and the Last Day…". The Prophet Muhammad (PBUH) built a mosque in Medina as his foremost act after emigrating from Makkah. Although women can pray with the congregation in the mosque, it is not as binding on them as men. In this context, the mosque remains a permanent feature of any Muslim community, irrespective of its material status: the richest or the poorest.

Mosques are differently managed in different countries and can be broadly categorised as state-run mosques and community-run mosques. In many Arab countries such as Egypt, Oman, Qatar, Saudi Arabia and the United Arab Emirates, mosques are state-run, built and maintained. Imams are appointed by the government. In the case of Pakistan, only important mosques such as the Badshahi Mosque in Lahore are state-run, managed by the *Auqaf* department. The vast majority of mosques are built and maintained by communities with imams appointed locally. This situation is the same as that in western countries where mosques are also owned and maintained by local communities. In state-run mosques, the state has control over sermons and on messages, an imam can deliver from mosques, whereas it has minimum control, if any, in the case of community-run mosques including those connected with urban centres. Local mosques attached to small village communities are not likely to have any strings attached. Each mosque is an independent entity with no hierarchical relationship with other mosques in the area.

2.5.1 *Is the Mosque an "Organisation" or an "Institution"?*

Other than those run by the government, civil society includes a variety of organisations and institutions founded by people themselves. It is an important question to consider whether the mosque is an "organisation" or an "institution" and what is the distinction between an "institution" and an "organisation" since both the terms are sometimes used interchangeably in a confusing manner. North (1991) suggests that institutions involve the human-devised constraints that structure political, economic and social interaction. They consist of both informal constraints (sanctions, taboos, customs, traditions and codes of conduct), and formal rules (constitutions, laws, property rights). On similar lines, Thomson (1995) argues that some institutions are also organisations, for example, public administration, political parties, the congregation of a mosque or a church, and the family. Such viewpoints show that the distinction between institutions and organisations blurred in development policy and practice, sometimes resulting in undermining the role of organisations. This has certainly been the argument of Leftwich and Sen (2010) who assert that "institutions are not self-generating or self-sustaining and they achieve little on their own". They conclude that it is the interaction of institutions with organisations, groups and individuals that shapes, determines and influences the development outcomes of institutions. As players of the game, organisations are a vehicle for the expression and articulation of collective interest in a formal or informal way. Both institutions and organisations can be formal, usually with written laws and regulations which can be enforced by third parties, and informal, usually with unwritten routines, norms, conventions and customs which are embedded in everyday practices and beliefs of life and its manifestations. For the same reason, both can be social, political and economic.

Civil society organisations comprise different types of voluntary associations including religious ones. Owing to the large number and complexity of these organisations, defining non-governmental organisations was dubbed as "mission impossible" by Martens (2002). Drawing on the work of Martens (2002), Berger (2003) attempts to provide an overarching definition of religious organisations:

> Organisations whose identity and mission are self-consciously derived from the teachings of one or more religious or spiritual traditions and which operate on a non-profit, independent, voluntary basis to promote and

realise collectively articulated ideas about the public good at the national or international level.

Recently, efforts have been made to define and determine the parameters of religious organisations to identify the ones of interest. Clarke (2006) classifies religious organisations into five faith-based categories: representative organisations or apex bodies, charitable or development organisations, socio-political organisations, missionary organisations that spread key faith messages beyond the faithful by seeking converts to it, and terrorist organisations which engage in illegal practices based on religious beliefs.

However, using North's (1991) definition, this book classifies the mosque as an institution since it has a set of rules, written and unwritten. Employing the analogy used by Leftwich and Sen (2010), a community and an imam are considered as players of a game in an organisational vehicle that operates through the institution of the mosque. The mosque articulates and governs the community's religious, social, cultural, spiritual, psychological, economic and political aspects of life through its organisational vehicle. The mosque is primarily a civil society and a community-based religious institution with overarching effects on the economic and political spheres of its actors. It is both a formal institution, since there are written injunctions regarding the role of the mosque in Islamic jurisprudence, and an informal one because the social character of the mosque in a specific rural community setting largely remains unwritten. It is this social character that is under scrutiny in this book and is shaped by several factors such as perception of communities and imams. Unlike historical and architectural mosques managed by the government, most mosques in Pakistan located in both urban and rural areas are built, maintained and run by communities themselves. Therefore, the government has no say in the affairs of a mosque as such. In this way, it is purely in the domain of the concerned community to determine the boundaries of the social character of their mosque.

The community and imam, a group and an individual, respectively, are two major players that shape and determine the role of the mosque as an institution and influence outcomes of development initiatives by other actors in disaster management. There are many features uniformly found across all the case study sites included in this study making a case for the mosque to be called both a formal and informal institution. These include facing Makkah (formal), being led by a male imam (formal), being chosen by the community (informal), gathering male Muslims five times a day through *Azaan* (formal), having female Muslims listening to Friday sermons (informal), involving children through the learning of the Holy Quran (informal), and the mosque being sustained by the concerned community without any government patronage or support (informal).

2.5.2 *Is the Mosque a "Faith" Institution or a "Religious" Institution?*

The difference in interpretation and usage of terms "religion" and "faith" has led to a difference between organisations and institutions attached to these terms. Bano and Nair (2007) after tracing the historical evolution of faith-based organisations in three countries of South Asia—Bangladesh, India and Pakistan—prefer the term "faith" over "religion". They contest that the term "faith" is broader than that of "religion", thus allowing flexibility to extend beyond major religions. Berger (2003) has used the term "religious non-governmental organisations" to conduct an exploratory analysis of 263 UN-affiliated organisations from different religions. Clarke (2007) has consistently used the term "faith" instead of "religion" or "religious" to denote faith-based organisations belonging to different religions.

This book prefers the term "religious" since it describes the mosque, a place of worship primarily, more strongly than the word "faith". In the academic literature, for example, in a study on the role of the mosque in local governance in Afghanistan, the mosque is referred to as a religious institution (Rahmani, 2006). This selection is useful because this research analyses the role of one religious entity, the mosque, in a Muslim society and therefore does not require a loose term like "faith", which appears more suitable for comparison among various religious institutions or organisations. The term "religious" is useful in directly denoting the mosque enabling this book to stay focussed and avoid confusion with other faith-based organisations. Overall, this book examines the role of the mosque in the 2005 earthquake disaster in the capacity of a religious institution.

2.6 Political Controversy About the Role of the Mosque

This study inquires whether the community-based religious institution of the mosque has a role to play in disaster management. However, it also acknowledges the political controversy, if not risk, implicit in this proposition. In the aftermath of the 9/11 attacks in the USA, the subject of the political instrumentalisation of Islam has received greater attention (Platteau, 2010). Mosques in Muslim-majority countries are known to have been used effectively by radical Muslim organisations such as Hamas in Palestine, Hizbullah in Lebanon, the Muslim Brotherhood in Egypt, Taliban in Afghanistan and Jamaat-ud-Dawah in Pakistan, to enhance their influence among the public (Gupta & Mundra, 2005). Jamaat-ud-Dawah had also a conspicuous presence in the relief operations after the 2005 earthquake (Qureshi, 2006). This was due to the ability of the organisation to provide immediate relief items than the government and other organisations working in the area.[7]

In Muslim-majority countries such as Pakistan, mosques having a pro-USA political stance—mainly against the armed struggle against the USA occupation of neighbouring Afghanistan—have been attacked by dissidents and worshippers killed (Usmani et al., 2010). In addition, mosques and attached seminaries are alleged to fan sectarian violence like the Shia-Sunni conflict in Pakistan (Haleem, 2003). Mosques in Muslim-minority countries, such as Europe, have been accused of fuelling hatred for the West and serve as recruiting grounds for potential terrorists. Instances include the bombings in Madrid (2004) and London (2005) (Dunn, 2001; Haddad & Balz, 2008).

Responding to concerns over the role of the mosque, western countries adopted a two-prong strategy: firstly, immediate measures such as increased surveillance of mosques and detention and deportation of alleged radical imams, and secondly, strategic measures to engage by investment in mosques and training imams to achieve better integration of Muslim communities into society (Innes, 2006). In this way, mosques

[7] Some parts of this section draw on my work during doctoral thesis titled "exploring the role of the mosque in dealing with disasters: A cases study of the 2005 earthquake in Pakistan" at Massey University, New Zealand available at http://hdl.handle.net/10179/4080.

have been instrumental in promoting the political and social integration of Muslim minorities in western countries such as the USA.

Likewise, mosques and imams have been deemed an important source for influencing public opinion in either Muslim-majority or -minority countries (Al-Astewani, 2021; Haqqani, 2005; Wong, 2010). If western countries with a minority Muslim population have chosen to engage with the mosque to address their issues (of national security and social integration), Muslim-majority countries would seem to have an even stronger reason to engage with the mosque for the religiously virtuous cause of securing lives and assets from disasters.

The COVID-19 pandemic offered yet another example where Imams and mosques engaged with the state to respond to public demands to the closure of mosques in Britain (Al-Astewani, 2021). The expert medical advice issued by the British Islamic Medical Association was widely accepted and the British Board of Imams and Scholars (BBSI) coordinated with both *Wifaqul Ulama* and the Mosques and Imams National Advisory Board when issuing its religious guidance. The religious leaders in the Muslim community forged their sectarian differences for the broader benefit of overall community welfare. Taking a step further, some of the mosques opened their premises for the COVID-19 vaccine to dispel false propaganda about the vaccine (BBC, 2021).

2.7 The Multi-faceted Role of the Mosque in Pakistan

The mosque has played a multi-faceted role in Pakistani society. People in Pakistan tend to trust religious institutions and donate generously to such an extent that mosques (and attached seminaries) impart basic literacy to one-third of the uneducated children of the country (Grare, 2007). As the role of the mosque in disaster management is understudied, this section explores the general role of the mosque in community development and acts of public welfare in different parts of Pakistan.[8]

[8] Some parts of this section draw on my work during doctoral thesis titled "exploring the role of the mosque in dealing with disasters: A cases study of the 2005 earthquake in Pakistan" at Massey University, New Zealand available at http://hdl.handle.net/10179/4080.

2.7.1 Water Conservation

Active civil society institutions such as mosques may become critical where state institutions lack the capacity and public faith to uphold the rule of the law. Shah et al. (2001) demonstrate a valuable achievement of mosques and religious schools in facilitating the implementation of the public policy about water conservation at the local level. A group of people who happened to socialise at the local mosque of a small town in Faisalabad (Pakistan) were concerned about the wastage of municipal and irrigation water supplies. This group of people (including the lead author, Shah) were not related to any formal or informal organisation. They studied the issue through a survey of 4,113 people and divided people into four groups based on their distance from the water reservoir as shown in Table 3.2.

Several people lodged complaints with the concerned government department about the water shortage and illegal installation of suction pumps by the people but all in vain: officials did not take any action. In this situation, Shah et al. (2001) approached the imams of the area concerned, to solicit them to discuss the issue of water conservation, in the framework of Islamic values, during Friday sermons. The lead author along with his team members also further educated imams on the issue of water conservation. The major focus was to spread awareness among the community on the rights of fellow Muslims and the importance of valuing precious resources, such as water, from the Islamic point of view. They wanted to mobilise and build on internal controls through promises of divine rewards and fear of punishments. The results of the first survey as compared to the second survey conducted two months later were:

It is evident from Table 2.1 that there was a significant decline in water wastage and the people in the 3rd and 4th group benefitted from this moral persuasion by the imams. Particularly, the tail-end users became the main beneficiaries. The authors concluded that the mosque and "Imams are more capable of reaching the public than water specialists" and that they had taken the policy from words to actions (Atallah et al., 2001). Religious institutions may help to reduce the trust gap between the state and civil society. Atallah et al. (2001) emphasised that such a partnership with state institutions would not only facilitate effective implementation of policy but instil a pro-environment behavioural change among community members which is fundamental to improvement in disaster management.

Table 2.1 Influence of the local mosque on water conservation

Groups	*Water shortage problem (percentage of people)*	
	Before survey	*After survey*
1st group (closer to water)	0	0
2nd group (close to water)	0	0
3rd group (away from water)	50	20
4th group (further away from water)	75	42

Source Author based on Shah et al. (2001)

2.7.2 Fighting Blindness

Sightsavers, a UK-based charity fighting blindness, has been working in Pakistan since 1985 (Sightsavers, 2021). The organisation has been engaging the community through local leadership including the Nazim (locally elected leader, equivalent of a Mayor), village health committees and the imam of the village mosque.

In 2008, Sightsavers chose to work in a small village, Killa Virkan, which was suffering from one of the highest levels of trachoma in the country, a type of local health disaster. Trachoma is a bacterial eye infection that is primarily caused by unsafe water and unhygienic conditions and leads to blindness. Syed Abbas Ali Shah, imam of the local mosque quoted, by the BBC (2008), inspired villagers to keep their streets, bazaars and homes clean by presenting them with the Prophetic narrations about cleanliness and hygiene. The organisation claims that the SAFE strategy (Surgery, Antibiotics, Face washing and Environmental hygiene) remained successful and the disease has been virtually eradicated from the village.

2.7.3 Fighting Poverty

Akhuwat, which means brotherhood, is a registered non-governmental organisation working to alleviate poverty with interest-free loans, developing entrepreneurial skills and capacity building since 2001 (Akhuwat, 2021). Over time, it has expanded its programmes and now operates

Pakistan's first fee-free boarding university near Lahore since 2018. It is based on the Islamic concept of Muslim brotherhood that refers to the earliest example of brotherhood established by the Prophet Muhammad (PBUH) between Meccan immigrants and citizens of Medina fourteen centuries ago. At that time, the citizens of Medina shared their wealth, businesses and houses with Meccan immigrants who were forced out of Makkah because of their conversion to a monotheistic religion, Islam, from polytheism, which was prevalent in the society.

The distinguishing feature of this organisation is its interest-free loans and use of religious places including mosques, churches and temples for loan disbursements.

The organisation's key lending methodology is to approach a community at the time of congregational prayer, either in a mosque or a church. People are given an introduction about the details of the lending programme and its costs. In the particular ambience of a religious place, borrowers feel an additional moral responsibility to return the loan. In addition, using the mosque as a space saves the organisation a huge operational cost in case it was to use different premises. Religious places ensure transparency, participation, and accountability and fostering goodwill among communities. Transparency is particularly an important aspect in countries including Pakistan where people do not fully trust governments due to prevalent corruption.

At the same time, the organisation believes in diversity by serving all members of society irrespective of their religion, caste, colour or gender. Transgender, still a taboo in Pakistani society, the organisation has special programmes offering cash and kind support through imparting technical and vocational skills and linking them with government AIDS control programme.

2.7.4 Campaigning on Birth Control

Religious influences have been recognised as a formidable challenge to the success of the government birth control policy in Pakistan (Abdur Cheema & Mehmood, 2018). The Ministry of Population Welfare has been working on population welfare and stabilisation issues but a formal acknowledgement of the role of religious community leaders/scholars/imam*s* as a part of civil society at the highest level became explicit in mid-2005. The Ministry organised an international conference on population and development inviting Islamic religious scholars/leaders from all over the world to Islamabad in May 2005 (Ministry of Population Welfare, 2006). The Islamabad declaration emphasising the role of Muslim religious scholars in population planning was issued. A year later, the Federal Minister informed the follow-up meeting of the Islamabad Declaration of religious leaders that:

> The awareness and training programme for local Ulama (religious scholars) has been strengthened to sensitize them on population issues for seeking their (public) support. I am glad to inform you that we have made remarkable progress in achieving this objective and *Inshallah* (with the support of Allah) very soon we will achieve the target of training 13000 Ulama all over Pakistan. Yet another important related development is the involvement of female Ulama in the programme activities who shall serve as a catalyst for behavioural change and community mobilization at the grassroots level.

This understanding and engagement continues as President of Pakistan Dr Arif Alvi, chaired a meeting of the population welfare department alongside Governor Punjab, Chaudhry Sarwar and Chief Minister Usman Buzdar stressed that ulema (religious scholars) should be engaged to raise awareness on issues like breastfeeding, nutrition, maternal nutrition and other related matters (*The Dawn*, 2020).

2.7.5 Fighting COVID-19 with Mosques

During the ongoing COVID-19 pandemic, the government of Pakistan and other national and international organisations collaborated with religious leaders to have their support in promoting healthcare behaviours. Given advances in the scientific understanding of disease transmission, officials recommended hand washing, sanitising, and most important, social distancing. With the onset and rise in disease spread, social distancing and restrictions on gatherings were extended to mosques in Pakistan. Over 260,045 religious leaders were engaged and mobilised to emphasise the importance of physical distancing and promoting key preventive messages building risk perception until August 15, 2020 (UNICEF, 2020). Though the Pakistan government collaborated with prominent religious leaders at the national level and had a consensus on a 20-point plan, this plan had limited acceptance and implementation at the community level. This strategy showed limited success as it did not involve ongoing and interactive engagement with a religious leader.

Violent protests were also witnessed in some places once Friday congregations were banned by the government under lockdown. A randomised control trial showed that when actively engaged through one-on-one persuasion, Imams can be mobilised to instruct congregants to take public health measures in their mosques (Vyborny, 2020).

The influence of religiosity on following COVID-19 health guidelines has gained significant attention. Health authorities insisted on restricting public movement through lockdowns on educational institutes, shopping malls and religious institutions. How these restrictions were perceived and implemented by faith communities gained the attention of academics. Studies have been conducted in developing and developed countries. A study in the USA shows that religiosity increases reactance and non-compliance of COVID-19 prevention health advisories (DeFranza et al., 2020).

2.8 Potential Role of Community-Based Religious Institutions Regarding "connect and Convince" Functions

The current Sendai Framework for Disaster Risk Reduction (2015–2030) and before Sendai Framework, the Hyogo Framework 2000–2015, identified the need for building resilience at all levels by promoting community engagement and networking (UNISDR, 2005, 2015). The Sendai Framework calls for the protection of religious places along with places of cultural heritage during disasters. The Hyogo Framework proposed a "connect and convince" typology to spread its message of saving lives and reducing the impact of disasters to all stakeholders including communities, governments, donors and experts. The connect function emphasises four dimensions—coordinate, campaign, inform and advocate—and the convince function also reinforces four aspects—organise, promote, encourage and provide—so that disaster management will achieve a disaster-safe future for all[9] (UNISDR, 2005).

All connect and convince functions may be effectively performed by a religious institution such as the mosque which has a central position

[9] Some parts of this section draw on my work during doctoral thesis titled "exploring the role of the mosque in dealing with disasters: A cases study of the 2005 earthquake in Pakistan" at Massey University, New Zealand available at http://hdl.handle.net/10179/4080.

in a Muslim community. Since outreach of the mosque message includes men, women and children five times a day (through the repeated call for prayers), it may connect—coordinate, campaign, advocate, inform and convince—organise, promote, encourage and become an important channel for promoting DRR activities and culture in Muslim communities.

Although each of these functions is important, coordination is essential for effective disaster management, but it is equally a challenging task. Coordination involves conducting particular tasks with mutually agreed upon cooperation and not just sharing information. However, there are limits to the effectiveness of disaster management and it should not be assumed to work out ideally as per principles. Despite knowing different social, economic and political factors, these factors may not be considered during implementation because of the complexity of a disaster situation.

Figure 2.1 shows the potential roles of the mosque regarding connect and convince functions. In this context, other community-based religious

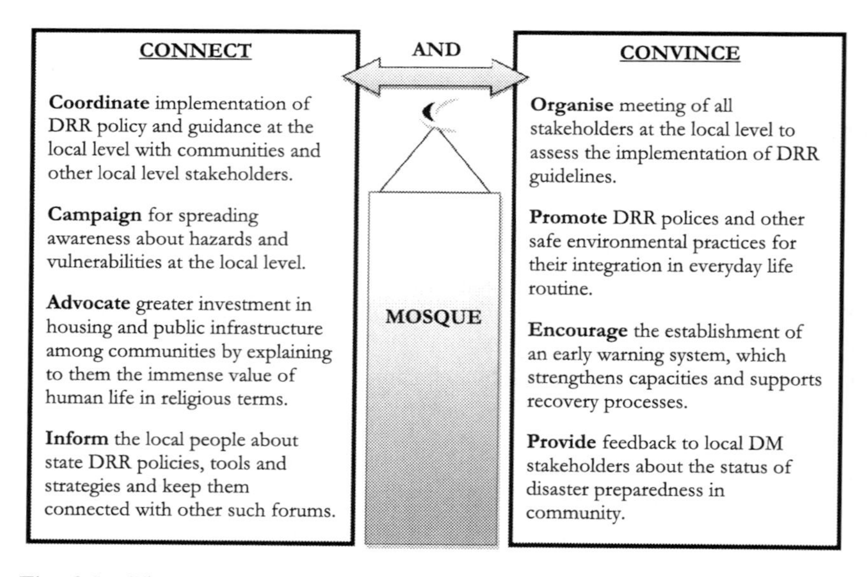

Fig. 2.1 The mosque and the UNISDR connect and convince functions (*Source* Author)

institutions such as churches, synagogues or temples may perform connect and convince functions permanently, if effectively engaged in disaster management.

2.9 Conclusion

This chapter illustrated the increasingly important role of community-based religious institutions in disaster management. While civil society began to receive attention with the emergence of the integrated approach to disasters, the role of religious institutions as a member of civil society remained neglected and undocumented. Although underreported in disaster and development literature, community-based religious institutions have been contributing to disaster preparedness, response, relief, reconstruction, rehabilitation and recovery. Since religious influences are usually strong at the local level, people's worldviews and perceptions about disasters are often religiously inspired.

With the beginning of the new millennium, religious and faith-based organisations received global attention, including the famous Jubilee 2000 campaign inspiring for debt relief for poor countries. Bilateral development organisations such as the DFID formally acknowledged the privileged position of the local religious institutions, which are based at a grassroots level and enjoy strong community support. Despite the conventional divide between religious and secular organisations, both began to find commonality in their objective of helping humanity and have started to forge closer links in the new millennium. In a rare move, DFID formed a strategic alliance against poverty with Muslim faith-driven organisations including Muslim Aid and Islamic Relief.

Religious institutions usually represent local and cultural sensitivity. It is increasingly difficult for international development and humanitarian organisations to be perceived as neutral, given the change in the global political climate since 9/11. In this context, ICRC suggests humanitarian organisations engage with religious leaders and institutions to dissuade any political impressions that may be attached to these organisations.

Religious institutions are an important determinant of culture. The shape and influence communities' worldviews, risk perception and attitudes towards disaster preparedness in faith-based societies. Risk perception is fundamental to behaviour towards disaster preparedness. If religious institutions promote fatalistic attitudes in communities, the disaster management policy of the state is not likely to achieve its desired level of safety and preparedness for the people. Nevertheless, religiously driven holistic worldviews of life may offer psychosocial healing, promote resilience and a kind of survival strategy for disaster victims.

This study categorises the mosque as a social, community-based and religious institution. This study explores the role of the mosque, as a key community-based religious institution where Islam prevails, for coping with disasters and reducing future disaster risk. Furthermore, this study also has relevance to the role of other community-based religious institutions such as churches, temples and synagogues in managing disasters.

Given the lack of literature on the role of the mosque in disasters, this chapter has explored the general developmental role of the mosque in Pakistani society to set the stage for the empirical investigation of the role of the mosque in the aftermath of the 2005 earthquake and during the ongoing COVID-19 pandemic. Mosques have been used for advocacy campaigns such as water conservation and the elimination of blindness. Imams have been found to have a significant influence on public opinion at the local level. The Ministry of Population Welfare has formally acknowledged the role of religious leaders including Ulama and imams as important public opinion makers and therefore engaged with them. Through training and engagement, imams as influential individuals and mosques as community institutions may contribute to improvement in the quality of life of a community by strengthening cohesion or promoting brotherhood and social safety networks. This improvement may become more valuable where the state cannot deliver due to a lack of resources or inefficiency.

Overall, the chapter has demonstrated that community-based religious institutions such as the mosque can influence the social, cultural, psychosocial, economic and political aspects of life. These aspects of the earthquake-affected communities' lives will be analysed during the process of data collection about the roles of mosques in the disaster management cycle, in the aftermath of the 2005 earthquake in Pakistan.

References

Abdel-Hady, Z. M. (2010). *The masjid, yesterday and today*. The Centre for International and Regional Studies, Georgetown University School of Foreign Service.

Akhuwat. (2021). *Akhuwat's story* (Akhuwat, Ed.). Akhuwat. http://www.akhuwat.org.pk

Al-Astewani, A. (2021). To open or close? COVID-19, mosques and the role of religious authority within the British Muslim community: A socio-legal analysis. *Religions, 12*(1), 1–26. https://doi.org/10.3390/rel12010011

Antonovsky, A. (1979). *Health, stress and coping*. Jossey-Bass Publishers.

Arutz Sheva. (2009). Joseph's era coins found in Egypt. *Arutz Sheva 7*. http://www.israelnationalnews.com/News/News.aspx/133601

Atallah, S., Khan, M. Z. A., & Malkawi, M. (2001). Water conservation through public awareness based on Islamic teachings in the Eastern Mediterranean region. In N. I. Faruqui, A. K. Biswas, & M. J. Bino (Eds.), *Water Management in Islam* (pp. 49–60). United Nations University Press.

Bankoff, G. (2004). The historical geography of disaster: Vulnerability and local knowledge. In G. Bankoff, G. Frerks, & D. Hilhorst (Eds.), *Mapping vulnerability: Disaster, development and people* (pp. 25–36). Earthscan.

Bano, M., & Nair, P. (2007). *Faith-based organizations in South Asia: Historical evolution, current status and nature of interaction with the state*. University of Birmingham.

BBC. (2008). *Clean water is essential for safeguarding eyes from blindness*. BBC URDU.Com. http://www.bbc.co.uk/urdu/multimedia/

BBC. (2021, January 21). *Birmingham mosque becomes UK's first to offer Covid vaccine*. https://www.bbc.com/news/uk-england-birmingham-55752056

Berger, J. (2003). Religious nongovernmental organizations: An exploratory analysis. *Voluntas: International Journal of Voluntary and Nonprofit Organizations, 14*(1), 15–39.

Berkes, F. (2008). *Sacred ecology* (2nd ed.). Taylor & Francis.

Bernstein, P. L. (1996). *Against the Gods: The remarkable story of risk*. Wiley.

Bonney, R., & Hussain, A. (2001). *Faith communities and the development agenda*. Centre for the History of Religious and Political Pluralism, University of Leicester. http://www.dfid.gov.uk/pubs/files/faithdevcomagenda.pdf

Brummitt, C. (2006). Indonesians see disasters as God's will. *The Christian Post*. http://www.christianpost.com/news/indonesians-see-disasters-as-god-s-will-1611/

Cairns, E. (2012). *Crises in a new world order: Challenging the humanitarian project*. Oxfam International.

Candland, C. (2000). Faith as social capital: Religion and community development in Southern Asia. *Policy Sciences, 33*(3), 355–374.

Cheema, A. R., Scheyvens, R., Glavovic, B., & Imran, M. (2014). Unnoticed but important: Revealing the hidden contribution of community-based religious institution of the mosque in disasters. *Natural Hazards*, 1–23. https://doi.org/10.1007/s11069-013-1008-0

Cheema, A., & Mehmood, A. (2018). Reproductive health services: 'Business in a Box' as a model social innovation. *Development in Practice*. https://doi.org/10.1080/09614524.2018.1541166

Chester, D. K. (2005). Theology and disaster studies: The need for dialogue. *Journal of Volcanology and Geothermal Research, 146*(4), 319–328. https://doi.org/10.1016/j.jvolgeores.2005.03.004

Chester, D. K., & Chester, O. K. (2010). The impact of eighteenth century earthquakes on the Algarve region, southern Portugal. *Geographical Journal, 176*(4), 350–370. https://doi.org/10.1111/j.1475-4959.2010.00367.x

Chester, D. K., Duncan, A. M., & Dibben, C. J. L. (2008). The importance of religion in shaping volcanic risk perception in Italy, with special reference to Vesuvius and Etna. *Journal of Volcanology and Geothermal Research, 172*(3–4), 216–228. https://doi.org/10.1016/j.jvolgeores.2007.12.009

Clarke, G. (2006). Faith matters: Faith-based organisations, civil society and international development. *Journal of International Development, 18*(6), 835–848. https://doi.org/10.1002/jid.1317

Clarke, G. (2007). Agents of transformation? Donors, faith-based organisations and international development. *Third World Quarterly, 28*(1), 77–96.

Coppola, D. P. (2007). *Introduction to international disaster management*. Butterworth-Heinemann.

Cottrell, A. (2006). Weathering the storm: Women's preparedness as a form of resilience to weather related hazards in Northern Australia. In D. Paton & D. Johnston (Eds.), *Disaster resilience: An integrated approach* (pp. 128–142). Charles C Thomas.

DeFranza, D., Lindow, M., Harrison, K., Mishra, A., & Mishra, H. (2020). Religion and reactance to COVID-19 mitigation guidelines. *American Psychologist*. https://doi.org/10.1037/amp0000717

Deneulin, S., & Zampini Davies, A. (2017). Engaging development and religion: Methodological groundings. *World Development, 99*, 110–121. https://doi.org/10.1016/J.WORLDDEV.2017.07.014

Deneulin, S., & Rakodi, C. (2011). Revisiting religion: Development studies thirty years on. *World Development, 39*(1), 45–54. https://doi.org/10.1016/j.worlddev.2010.05.007

Department for International Development. (2001). *Target 2015 halving world poverty: A shared vision of reducing world poverty – British Government and British Muslim charities working to realise the common good*. Department for International Development (DFID). http://www.dfid.gov.uk/pubs/files/2015-muslim.pdf

Dunn, K. M. (2001). Representations of Islam in the politics of mosque development in Sydney. *Tijdschrift Voor Economische En Sociale Geografie, 92*(3), 291–308. https://doi.org/10.1111/1467-9663.00158

Feener, R. M., & Fountain, P. (2018). Religion in the age of development. *Religions, 382*(9), 1–23. https://doi.org/10.1017/CBO9780511621475

Fozdar, F., & Roberts, K. (2010). Islam for fire fighters - a case study on an education programme for emergency services. *The Australian Journal of Emergency Management, 25*(1), 47–53. http://www.ema.gov.au/www/emaweb/rwpattach.nsf/VAP/%288AB0BDE05570AAD0EF9C283AA8F533E3%29~Roberts+&+Fozdar.pdf/$file/Roberts+&+Fozdar.pdf

Gaillard, J. C., & Texier, P. (2010). Religions, natural hazards, and disasters: An introduction. *Religion, 40*(2), 81–84. http://www.sciencedirect.com/science/article/B6WWN-4YGHK8H-1/2/91b2bc41f3be1b58ac1ec07066f66853

Gianisa, A., & Le De, L. (2018). The role of religious beliefs and practices in disaster: The case study of 2009 earthquake in Padang city, Indonesia. *Disaster Prevention and Management: An International Journal, 27*(1), 74–86. https://doi.org/10.1108/DPM-10-2017-0238

Gibson, M. (2006). *Order from chaos: Responding to traumatic events*. The Policy Press.

Gingerich, T. R., Moore, D. L., Brodrick, R., & Beriont, C. (2017). *Local humanitarian leadership and religious literacy: Engaging with religion, faith, and faith actors*. https://doi.org/10.21201/2017.9422

Grandjean, D., Rendu, A.-C., MacNamee, T., & Scherer, K. R. (2008). The wrath of the gods: Appraising the meaning of disaster. *Social Science Information, 47*(2), 187–204. https://doi.org/10.1177/0539018408089078

Grare, F. (2007). The evolution of sectarian conflicts in Pakistan and the ever-changing face of Islamic violence. *South Asia-Journal of South Asian Studies, 30*(1), 127–143. https://doi.org/10.1080/00856400701264068

Griffiths, V. F., Bull, J. W., Baker, J., Infield, M., Roe, D., Nalwanga, D., Byaruhanga, A., & Milner-Gulland, E. J. (2020). Incorporating local nature-based cultural values into biodiversity No Net Loss strategies. *World Development, 128*, 104858. https://doi.org/10.1016/j.worlddev.2019.104858

Guarnacci, U. (2016). Joining the dots: Social networks and community resilience in post-conflict, post-disaster Indonesia. *International Journal of Disaster Risk Reduction, 16*, 180–191. https://doi.org/10.1016/j.ijdrr.2016.03.001

Gupta, D. K., & Mundra, K. (2005). Suicide bombing as a strategic weapon: An empirical investigation of Hamas and Islamic Jihad. *Terrorism and Political Violence, 17*(4), 573–598. https://doi.org/10.1080/09546550500189895

Ha, K. M. (2015). The role of religious beliefs and institutions in disaster management: A case study. *Religions, 6*(4), 1314–1329. https://doi.org/10.3390/rel6041314

Haddad, Y. Y., & Balz, M. J. (2008). Taming the Imams: European governments and Islamic preachers since 9/11. *Islam and Christian–Muslim Relations, 19*(2), 215–235. http://www.informaworld.com/10.1080/09596410801923980

Haleem, I. (2003). Ethnic and sectarian violence and the propensity towards praetorianism in Pakistan. *Third World Quarterly, 24*(3), 463–477. https://doi.org/10.1080/0143659032000084410

Haqqani, H. (2005). *Pakistan: Between mosque and military*. United Book Press.

Haque, N. ul. (2009). How to solve Pakistan's problems. *Open Democracy*. http://www.opendemocracy.net/article/how-to-solve-pakistan-s-problem

Hirono, T., & Blake, M. E. (2017). The role of religious leaders in the restoration of hope following natural disasters. *SAGE Open, 7*(2). https://doi.org/10.1177/2158244017707003

Holmgaard, S. B. (2019). The role of religion in local perceptions of disasters: The case of post-tsunami religious and social change in Samoa. *Environmental Hazards, 18*(4), 311–325. https://doi.org/10.1080/17477891.2018.1546664

Homan, J. (2003). The social construction of natural disaster: Egypt and the UK. In M. Pelling (Ed.), *Natural disasters and development in a globalizing world* (pp. 141–156). Routledge.

Hutton, D., & Haque, C. E. (2003). Patterns of coping and adaptation among erosion-induced displacees in Bangladesh: Implications for hazard analysis and mitigation. *Natural Hazards, 29*(3), 405–421. https://doi.org/10.1023/A:1024723228041

Innes, M. (2006). Policing uncertainty: Countering terror through community intelligence and democratic policing. *The ANNALS of the American Academy of Political and Social Science, 605*(1), 222.

Joakim, E. P., & White, R. S. (2015). Exploring the impact of religious beliefs, leadership, and networks on response and recovery of disaster-affected populations: A case study from Indonesia. *Journal of Contemporary Religion, 30*(2). https://doi.org/10.1080/13537903.2015.1025538

Kennedy, F. (2001). Sicilians pray as technology fails to stop lava. *The Independent*. http://www.encyclopedia.com/doc/1P2-5177590.html

Khan, W. (2011). *Say what we should do?* BBC Urdu.Com. http://www.bbc.co.uk/urdu/pakistan/2011/09/110928_naukot_wusat_ar.shtml

Laditka, S. B., Murray, L. M., & Laditka, J. N. (2010). In the eye of the storm: Resilience and vulnerability among African American women in the wake of Hurricane Katrina. *Health Care for Women International, 31*(11), 1013–1027. https://doi.org/10.1080/07399332.2010.508294

Leach, M., MacGregor, H., Scoones, I., & Wilkinson, A. (2020). Post-pandemic transformations: How and why COVID-19 requires us to rethink development. *World Development*, 105233. https://doi.org/10.1016/j.worlddev.2020.105233

Leftwich, A., & Sen, K. (2010). *Beyond institutions. Institutions and organizations in the politics and economics of poverty reduction - a thematic synthesis of research evidence*. DFID-funded Research Programme Consortium on improving Institutions for Pro-Poor Growth (IPPG).

Mahr, M. A. D. (2005). *The role of the mosque in the building of the society* (in Urdu). An-Noor Publications.

Malthus, T. R. (1958). *An essay on population* (M. P. Fogarty, Ed., Vol. 1). Dent.

Martens, K. (2002). Mission impossible? Defining nongovernmental organizations. *Voluntas: International Journal of Voluntary and Non-Profit Organizations, 13*(3), 271–285.

McGeehan, K. M., & Baker, C. K. (2017). Religious narratives and their implications for disaster risk reduction. *Disasters, 41*(2), 258–281. https://doi.org/10.1111/disa.12200

McGregor, A. (2010). Geographies of religion and development: Rebuilding sacred spaces in Aceh, Indonesia, after the tsunami. *Environment and Planning A, 42*(3), 729–746.

Ministry of Population Welfare. (2006). *Proceedings of follow-up meeting of the council of participating countries: Islamabad declaration on population and development* (M. of P. Welfare, Ed.). http://www.mopw.gov.pk/ulamafollowup/fupindex.html

Ngin, C., Grayman, J. H., Neef, A., & Sanunsilp, N. (2020). The role of faith-based institutions in urban disaster risk reduction for immigrant communities. *Natural Hazards, 103*(1), 299–316. https://doi.org/10.1007/s11069-020-03988-9

Ni Raghallaigh, M., & Gilligan, R. (2010). Active survival in the lives of unaccompanied minors: coping strategies, resilience, and the relevance of religion. *Child & Family Social Work, 15*(2), 226–237. https://doi.org/10.1111/j.1365-2206.2009.00663.x

North, D. (1991). Institutions. *Journal of Economic Perspectives, 5*(1), 97–112.

Oliver-Smith, A. (2004). Theorizing vulnerability in a globalized world: A political ecological perspective. In G. Bankoff, G. Frerks, & D. Hilhorst (Eds.), *Mapping vulnerability: Disaster, development and people* (pp. 1–10). Earthscan.

Paradise, T. R. (2005). Perception of earthquake risk in Agadir, Morocco: A case study from a Muslim community. *Global Environmental Change Part B: Environmental Hazards, 6*(3), 167–180. http://www.sciencedirect.com/science/article/B6VPC-4KPP4DD-1/2/4b72955cce74574bfb79d3b3e1c778c0

Platt, R. H. (1998). Planning and land use adjustments in historical perspective. In R. J. Burby (Ed.), *Cooperating with nature: Confronting natural hazards with land-use planning for sustainable communities* (pp. 29–56). Joseph Henry.

Platteau, J. P. (2010). Political instrumentalization of Islam and the risk of obscurantist deadlock. *World Development, 39*(2), 243–260.

Pooley, J. A., Cohen, L., & O'Connor, M. (2006). Links between community and individual resilience: Evidence from cyclone affected communities in

North West Australia. In D. Paton & D. Johnston (Eds.), *Disaster resilience: An integrated approach* (pp. 161–173). Charles C Thomas.

Quarantelli, E. L. (2009). *The earliest interest in disasters and crises and the early social science studies of disasters as seen in a sociology of knowledge perspective* (Working Paper 91). Disaster Research Center, University of Delaware.

Quarantelli, E. L., Lagadec, P., & Boin, A. (2007). A heuristic approach to future disasters and crises: New, old, and in-between types. In H. Rodriguez, E. L. Quarantelli, & R. R. Dynes (Eds.), *Handbooks of disaster research* (pp. 16–41). Springer.

Qureshi, J. (2006). Earthquake jihad: The role of jihadis and Islamist groups after the October 2005 earthquake. *Humanitarian Exchange, June*(34), 40. https://odihpn.org/wp-content/uploads/2006/07/humanitarianexchange034.pdf

Rahmani, A. I. (2006). *The role of religious institutions in community governance affairs: How are communities governed beyond the district level?* Centre for Policy Studies, Central European University. http://pdc.ceu.hu/archive/00002849/01/rahmani.pdf

Sain, V. (2008). Jodhpur Muslims help temple stampede victims. *Hindustan Times*. http://www.hindustantimes.com/StoryPage/FullcoverageStoryPage.aspx?id=0bbee88b-fff7-424e-be94-479a78b51c5bTemplesofdoom_Special&&Headline=Jodhpur+Muslims+help+temple+stampede+victims

Schmuck, H. (2000). "An act of Allah": Religious explanations for floods in Bangladesh as survival strategy. *International Journal of Mass Emergencies and Disasters, 18*(1), 85–96.

Shah, S. M. S., Baig, M. A., Khan, A. A., & Malkawi, M. (2001). Water conservation through community institutions in Pakistan: Mosques and religious schools. In N. I. Faruqui, A. K. Biswas, & M. J. Bino (Eds.), *Water management in Islam* (pp. 61–67). United Nations University Press.

Sheikhi, R. A., Seyedin, H., Qanizadeh, G., & Jahangiri, K. (2020). Role of religious institutions in disaster risk management: A systematic review. *Disaster Medicine and Public Health Preparedness*. https://doi.org/10.1017/dmp.2019.145

Shore, B. (2002). Taking culture seriously. *Human Development, 45*(4), 228–266.

Sightsavers. (2021). *Our work in Pakistan*. http://www.sightsavers.org/whatwedo/ourworkintheworld/asia/pakistan/world6206.html

Stern, G. (2007). *Can God intervene?: How religion explains natural disasters*. Praeger Publishers.

Suhail, R. (2010). *Temples are better than camps*. BBC Urdu.Com. http://www.bbc.co.uk/urdu/pakistan/2010/09/100902_flood_hindu_temple.shtml

Suri, K. (2018). Understanding historical, cultural and religious frameworks of mountain communities and disasters in Nubra valley of Ladakh. *International*

Journal of Disaster Risk Reduction, 31, 504–513. https://doi.org/10.1016/J.IJDRR.2018.06.004

The Dawn. (2020, November 17). President for engaging ulema in population control. https://www.dawn.com/news/1590804/president-for-engaging-ulema-in-population-control

Thomson, J. (1995). *Community institutions and the governance of local woodstocks in the context of Mali's democratic transition*. 38th Annual Meeting of the African Studies Association, November 3–6. http://dlc.dlib.indiana.edu/archive/00002673/01/Community_Institutions.pdf

Twigg, J. (2004). Good practice review: Disaster risk reduction, mitigation and preparedness in development and emergency programming. In *Disaster risk reduction: Mitigation and preparedness in development and emergency programming*. Overseas Development Institute. http://www.cababstractsplus.org/google/abstract.asp?AcNo=20043073047

UNICEF. (2017). *Civil society partnerships: Framework for engagement*. https://www.unicef.org/about/partnerships/index_60134.html

UNICEF. (2020). Pakistan COVID-19 Situation Report No. 15. 15, 1–14. https://www.unicef.org/media/78396/file/Pakistan-COVID19-SitRep-15-August-2020.pdf. 15. Accessed 7 September 2020.

UNISDR. (2005). *Hyogo framework for action 2005–2015: Building the resilience of nations and communities to disasters*. In World Conference on Disaster Reduction, 18–22 January 2005, Kobe, Hyogo, Japan (Issue A/CONF.206/6). United Nations International Strategy for Disaster Reduction (UNISDR).

UNISDR. (2015). *Sendai framework for disaster risk reduction*. UNISDR. https://www.preventionweb.net/files/43291_sendaiframeworkfordrren.pdf

Usmani, Z. U., Imana, E. Y., & Kirk, D. (2010). Escaping death - Geometrical recommendations for high value targets. In T. Sobh (Ed.), *Innovations and advances in computer sciences and engineering* (pp. 503–508). Springer-Verlag Berlin. https://doi.org/10.1007/978-90-481-3658-2_88

Vaught, S. (2009). An "Act of God": Race, religion, and policy in the wake of Hurricane Katrina. *Souls, 11*(4), 408–421. https://doi.org/10.1080/10999940903417276

Vyborny, K. (2020). *Persuasion and public health: Evidence from an experiment with religious leaders during COVID-19 in Pakistan* (IGC Working Paper, 1–21). https://www.dropbox.com/s/eb04gvie3hp19kd/Imams_live_web.pdf?dl=0

Wilkinson, O., Duff, J., Nam, S., Trotta, S., & Goodwin, E. (2020). *COVID 19: Practising our faith safely during a Pandemic - Adapting how we gather together, pray and practise*. Faith and Positive Change for Children, Families and Communities (FPCC). https://jliflc.com/covid/

Wisner, B. (2010). Untapped potential of the world's religious communities for disaster reduction in an age of accelerated climate change: An epilogue & prologue. *Religion, 40*(2), 128–131. http://www.sciencedirect.com/science/article/B6WWN-4YCNJPK-3/2/cf7680b3eda1132b8ff10f38a8690655
Wong, L. P. (2010). Information needs, preferred educational messages and channel of delivery, and opinion on strategies to promote organ donation: A multicultural perspective. *Singapore Medical Journal, 51*(10), 790–795.

CHAPTER 3

Disaster Management in Pakistan

3.1 Introduction

This chapter aims at providing an understanding of the key issues of disaster management policies and structures in the Pakistani context. The thematic arrangement of this chapter revolves around the roles of the key disaster management actors including the government, the private sector and civil society in the country. Disaster management initiatives are traced from the pre-independence period, and the British government's response-driven approach to the 1935 Quetta earthquake is examined to provide the background for disaster policies in the post-independence period.

Since Pakistan's disaster management policies were focussed solely on flood control before the 2005 earthquake, the analysis of disaster management policy and structure in Pakistan is divided into pre-and post-2005 earthquake periods. The pre-2005 earthquake disaster management policies are analysed through the assessment of ten 5-year developmental plans, from 1955 to 2010. The analysis of disaster management structures involves a review of the roles of 13 government ministries and departments related to flood-centric disaster preparations in the pre-2005 earthquake setting, based on primary and secondary sources.

A. R. Cheema, *The Role of Mosque in Building Resilient Communities*, Islam and Global Studies,
https://doi.org/10.1007/978-981-16-7600-0_3

After explaining the National Disaster Risk Management Framework (NDRMF), this chapter provides an analysis of the roles of 34 government ministries by comparing the pre-and post-2005 earthquake disaster management structure and policy, in the context of the disaster management cycle. During this analysis, this chapter highlights the issues of institutional overlap, conflict, ad hocery and the marginal role of civil society and community-based institutions in the formal disaster management structure. The analysis is enriched by incorporating interview excerpts from the two periods of fieldwork in 2009 and 2010. An upshot of the changes during and after 2010 including National Disaster Management Act 2010, National Disaster Risk Reduction Policy 2013, National Disaster Management Plan 2012, National Policy Guidelines on Vulnerable Groups in Disasters 2014, Host Nation Support Guidelines 2018 and National Disaster Response Plan 2019 is also included in this chapter. The roles of the mosque as a community institution and the imam as an opinion-maker are also introduced here as important stakeholders in disaster management at the community level.

3.2 Approaches to Disasters Since the 1935 Quetta Earthquake

During colonial rule, the British government set up a contingency-oriented infrastructure in the sub-continent for responding to any disaster. An example of a contingency approach (looking only at the immediate needs of response and relief) to disasters can be seen from the 1935 Quetta earthquake, which occurred during British rule in India. Before the independence of India and Pakistan in 1947, the British devised a contingency approach in the sub-continent to respond to disasters. Quetta is now the provincial capital of Balochistan which is the largest province of Pakistan in terms of area. An earthquake struck the people of Quetta in the morning of May 31, 1935, destroying the whole city and claiming 30,000 lives (Skrine, 1936; *The Times*, 1935b). A train full of doctors and nurses was sent from Karachi to Quetta (*The Times*, 1935a). The government was satisfied that substantial food supplies were available in Quetta and there was no fear of starvation. As a matter of

Table 3.1 Historical record of past earthquakes

Year	*Location*	*Magnitude*	*Deaths*	*Losses (Rs in Million)*
October 2015	KP, Punjab, AJ&K and GB	8.1	280	98,069 houses & 479 schools
September 2013	Awaran	7.7	376	6842 houses
October 2008	Ziarat	6.4	160	5943 houses
October 2005	KP&AJK	7.6	73,338	208,091
December 1974	Northern Area	7.4	5300	4400 houses
November 1945	Makran Coast	8.3	4,000	–
May 1835	Quetta	7.7	60,000	–

Source National Disaster Response Plan, NDMA (2019)

great urgency, martial law[1] was declared, and it was mainly a military-driven response and relief set up under the command of the army to help the Indian sufferers (Callisthenes, 1935). About 7,000 troops from the Quetta garrison joined the response and relief operation[2] (*The Times*, 1935a).

After independence in 1947, Pakistan could not free itself from the British colonial legacy of a contingency approach to natural hazards. Pakistan's institutional framework of disaster management has been predominantly reactionary, with limited focus only on response and relief, and to deal only with one hazard—flooding which has indeed been the most recurrent disaster to affect the largest number of people in the country since its establishment.

Earthquakes, though less frequent than floods, remained the deadliest events in the country's history. Since 1935, excluding the 2005 earthquake, earthquakes have killed 73,338 people in Pakistan (National Disaster Management Authority, 2019). Table 3.1 shows a historical record of the loss of lives and houses that occurred due to earthquakes.

However, it was not until the 2005 earthquake that attention was drawn to the rehabilitation of the earthquake-affectees (National Disaster

[1] Direct control of the country by the armed forces (Macmillan Dictionary, 2012).

[2] Some parts of this section draw on my work during doctoral thesis titled "exploring the role of the mosque in dealing with disasters: A cases study of the 2005 earthquake in Pakistan" at Massey University, New Zealand available at http://hdl.handle.net/10179/4080.

Management Authority, 2007). Thus, a clear paradigm shift occurred in the disaster management institutional policy and structure from a flood-centric to an integrated and multi-hazard approach to natural hazards in Pakistan. The discussion of disaster planning and implementation is therefore divided into pre-and post-2005 earthquake timeframes.

3.3 Pre-2005 Disaster Management Policies

In general, Pakistan's disaster management focus has been limited to counter-flooding policies. This was due to a lack of awareness concerning other disasters and the high frequency of floods. As a matter of legal arrangement and legislation, "The West Pakistan National Calamities (Prevention and Relief) Act 1958", generally called "The Calamities Act 1958", set parameters for the conduct of the state during natural hazardous events. This act, however, concentrated on the provision of response and relief to the affected communities. No amendments were made in the Calamities Act 1958 and a crisis management style towards disasters remained dominant in the country, as shown in Table 3.2, in the 5-year development plans for decades to come.[3]

The Medium-Term Development Framework, covering the period from 2005 to 2010, suggested a complete diversion from the flood-centred legacy of disaster to a multi-hazard disaster approach. It was launched in May 2005 (Planning Commission of Pakistan, 2010). For the first time, the subject of disaster preparedness and risk mitigation received full attention in the national planning corridors of power. What was behind this sudden change of approach to disasters in early 2005 in Pakistan? It was the December 26, 2004, Boxing Day Asian tsunami. Having seen the tsunami, world attention was drawn to the need to prepare for reduction or losses from natural hazardous events, and Pakistan was no exception. An international conference was held in January 2005, and the Hyogo Framework for disaster risk reduction was agreed upon by world leaders (UNISDR, 2005a). International development organisations such as the United Nations Development Programme (UNDP) quickly incorporated lessons from this destructive

[3] Some parts of this section draw on my work during doctoral thesis titled "exploring the role of the mosque in dealing with disasters: A cases study of the 2005 earthquake in Pakistan" at Massey University, New Zealand available at http://hdl.handle.net/10179/4080.

Table 3.2 Disaster management policies and related major events up until the 2005 earthquake

Five-year plans	*Disaster management policies, plans and major events*
First 5-year plan 1955–1960	The National Calamities Act 1958 passed because of recurrent flooding in East Pakistan. The Act is limited to response and relief. The focus on counter-flooding measures did not include flash flooding; measures restricted to river floods only
Second 5-year plan 1960–1965	Nothing new from the first 5-year plan except an increase in budget allocation for flood control measures
Third 5-year plan 1965–1970	Continued sole focus on river flooding control besides adding measures for enhancement of flood protection to increase the area under cultivation
Fourth 5-year plan 1970–1975	The contingency approach prevails. A cyclone hits East Pakistan and an Emergency Relief Cell (ERC) is set up at the federal level. Government must prepare an elaborate flood control programme with the technical support of the World Bank because of the political crisis in the country. East Pakistan becomes Bangladesh in 1971. Floods hit Pakistan in 1973 and 1976
No plan period 1971–1976	Due to the political crisis, the government falls back on annual planning—no 5-year plan period from 1971 to 1976. The federal government nationalises several private assets and many provincial functions such as canal rehabilitation and fertiliser subsidy. The National Calamities Act 1958 is re-adopted as the West Pakistan Calamities Act. The Act remains limited to response and relief
Fifth 5-year plan 1978–1983	A flood control policy is further centralised with the establishment of the Federal Flood Commission in 1977. The role of provincial and district governments is further reduced in local disaster risk and hazard mitigation planning
Sixth 5-year plan 1983–1988	Technocratic tendencies hold with the extension of irrigation and drainage systems. Absence of grassroots participation of the affected communities

(continued)

Table 3.2 (continued)

Five-year plans	*Disaster management policies, plans and major events*
Seventh 5-year plan 1988–1993	Structural measures such as the building of added storage capacity to store floodwaters and enhancing flood forecasting and flood warning system dominate the disaster policy horizon
Eighth 5-year plan 1993–1998	Focus on canal lining, remodelling and use of floodwater for land recharging. In addition, some non-structural measures, such as the promotion of water resources research in universities
Ninth 5-year plan 1998–2003	Flood control measures continued as in the earlier plans. However, the plan abandoned given extraordinary circumstances due to the 9/11 events and Pakistan's new role in the war on terror
Medium-Term Development Framework 2005–2010	Shift from flood-centred policy to a multi-hazard approach. UNDP Pakistan provided technical support and incorporated lessons learnt from the Boxing Day tsunami on December 26, 2004
Vision 2030	Poverty alleviation through control over natural calamities such as floods, droughts or human-induced events such as wars, and through the introduction of agriculture insurance against drought

Source Author

event. UNDP Pakistan functioned as a technical adviser to the government and reflected the latest learning from the tsunami during the preparation of the Medium-Term Development Framework. The UN had already reviewed the status of disaster management policies and the institutional structure of Pakistan during the preparation of the country report for the World Congress on Disaster Reduction (UNISDR, 2005b). The Medium-Term Development Framework referred to the December 2004 tsunami and suggested an all-inclusive review of the existing capacity of the public sector and civil society organisations for disaster preparedness and management. UNDP also aided provinces to prepare proposals for setting up disaster management structures. As a result, comprehensive disaster management plans are prepared for Punjab, Sindh, Balochistan, Northern Areas, and Azad Jammu and Kashmir.

In May 2005, the Medium-Term Development Framework proposed a National Disaster Management Authority (NDMA) at the federal level to design, organise, coordinate and implement disaster relief, rehabilitation, preparedness, early warning system and risk-reduction measures for all hazards across the country, but the October 8, 2005 earthquake still took the government by surprise.

3.4 Pre-2005 Disaster Management Structure

Given Pakistan's sole focus on flood control, the country developed a loosely organised, though not sufficient, control and command system for dealing with floods. It is therefore pertinent to examine the role of the different institutional structures that were in place or raised over time mainly to respond to the problem of river flooding before the new institutional mechanisms were put in place after the 2005 earthquake.

3.4.1 Government

Before 2005, there were about 27 federal and provincial institutions related to disaster response and relief. Since there was no coherent policy or central authority for dealing comprehensively and systematically with all disasters, no government plan elaborated clear demarcation of responsibilities of different government organisations in an integrated manner at different phases of a disaster or delineated their relationships. Table 3.3 shows disaster-related federal ministries and provincial departments in the pre-2005 earthquake disaster management structure.

Table 3.3 Disaster-related federal ministries and provincial departments in the pre-2005 earthquake disaster management structure

Ministry	*Department*	*Brief history, roles and responsibilities in disaster management*
Interior	Civil Defence Department	Established in 1951 at federal, provincial and district levels to ensure peace by preparing people in case of foreign country aggression. In 1993, emergency preparation, first aid, response and relief for all kind of hazards were included in its mandate
	Emergency Relief Cell (ERC)	Established in 1971 at the federal level to deal with the emergency in the aftermath of the cyclone in East Pakistan. Its job is to stockpile goods and relief items and coordinate with provincial relief departments. It ran as an emergency control room to liaise with other departments
	National Crisis Management Cell (NCMC)	Established in July 1999 under the Anti-Terrorist Act at the federal and provincial levels to deal with any emergency resulting from human or natural hazards
Water and Power	Water and Power Development Authority	Established in 1958, reservoir management and collection of rainfall data through its telemetric rain-gauge stations at separate locations across the Indus River System. Also runs a seismic observatory at Tarbela dam since 1974, which led to the detection of a new seismic zone—Indus Kohistan—the zone involving the 2005 earthquake
	The Indus River Commission	Established in 1960 after the signing of the Indus Waters Treaty, the Commission gathers data on river flow and rainfall in the catchment areas of Pakistan's western rivers flowing upstream from India
	Federal Flood Commission	Established in 1977 to have effective control of floods and to reduce flood losses. It prepares flood protection plans for the whole country
	Dams Safety Council	Established in 1987 to check dams' safety under federal and provincial governments and to coordinate with the Federal Flood Commission on large dams; based in Paris

Ministry	*Department*	*Brief history, roles and responsibilities in disaster management*
Defence	Frontier Works Organisation	Established during the construction of Karakoram Highway 1966–1978. Run by the army, it has the hi-tech logistic capability to unblock roads by removing landslides in hilly terrains
	Armed Forces	Pakistan Army, Air Force and Navy play leading roles in response, relief and evacuation. Different army organisations like Army Aviation Corps, Army Medical Corps and Army Engineers Corps have dominant roles in the early phases of a disaster
	Pakistan Meteorological Department	A key institution that collects and analyses rainfall data and shares information relating to weather and geophysical phenomena with aims of traffic safety in air, land and sea and therefore reduction in disasters and disaster losses. Shares information on flooding during the monsoon season, June 15 to September 30, with all provinces
	Flood Forecasting Division	Meant to collect, analyse and prepare a flood forecast and warning, as necessary. It is a subsidiary organisation of the Pakistan Meteorological Department
Cabinet Division	Planning Commission of Pakistan	Established in 1958 for strategic planning through preparation of national development plans with regular intervals. Its functions comprise allocation of resources including disaster management
	Space and Under Atmosphere Research Centre	Established in 1981 as a commission at the federal level. It conducts studies and projects on satellite remote sensing for hazard mapping, resource surveying and environmental monitoring to obtain information about impending disasters
Departments		*Provincial Departments and their roles*

(continued)

Table 3.3 (continued)

Ministry Department	*Brief history, roles and responsibilities in disaster management*
Planning and Development	Key planning body in each provincial government. Not directly involved in disaster risk planning but indirectly related since it formulates short-term and long-term provincial development plans
Irrigation	Undertake planning, designing and maintenance of flood protection work under the supervision of the Federal Flood Commission
Provincial Crisis Management Cell	Monitor and respond to any emergency, particularly disasters like terrorist activity; works under auspices of National Crisis Management Cell
Police	Present at the grassroots level, keep law and order during a disaster situation, disseminate flood warnings and help in search and rescue
Relief	Coordinate at the provincial level among several actors including federal, provincial and district governments and the affected community in a disaster. Interact with district governments to establish flood relief centres at district and tehsil levels. These departments usually worked under the Board of Revenue
Health	Support response and relief efforts by providing treatment to the injured. Declare emergency in hospitals in disaster situations and organise medical teams at a disaster location
Agriculture and Livestock	Reduce loss to livestock and agricultural land and help in the recovery of the same after a disaster by providing subsidised agriculture inputs like seed and fertilisers
Communication and Works	Responsible at the provincial level for maintenance and protection of communication network and infrastructure such as roads and bridges before and after a disaster
Food	The stockpile of food items and organisation of ration depots at the affected places to cater for the basic food requirements of the affected people
Punjab Emergency Service—Rescue 1122	Responsible for first call response to all emergencies. Established October 14, 2004, as a pilot project in Lahore (capital of the Punjab province) and now expanded to other districts of Punjab and provinces

Source Author

Table 3.3 summarises the Pakistani federal, provincial and district institutions involved in disaster management during different phases of a disaster before the 2005 earthquake. The table classifies these roles in the three phases—response, relief and preparedness—since the government ignored long-term rehabilitation plans. Some of these organisations would have an overlapping role in more than one phase of a disaster. For example, the Communication and Works Department mainly works during relief at different tasks such as restoring affected roads, but it would join Pakistan Army Engineers Corps to restore critical infrastructure during the response phase of a disaster. However, the roles of these institutions have been classified according to the focus of their interests and activities based on pre-2005 earthquake disaster management institutional policies and structures as identified in different reports, documents and interviews during the fieldwork.

The list of ministries and organisations produced in Table 3.4 includes major government organisations related to flood management and control. There were only four federal ministries (Interior, Defence, Cabinet and, Water and Power) focussed on flood control and management in the pre-2005 earthquake policy and structure for disaster loss reduction. Keeping in mind the restricted focus of the government on counter-flooding measures, ten[4] government departments were working for the response, including six at the federal level and four at the provincial level. Since the relief phase was the focus of the government, there were 14 organisations, including six at federal and eight at the provincial level. Investment in preparedness measures was not a priority for the federal government, and even less so at provincial and district levels. There were 13 government organisations expected to prepare plans for flood-centric disasters, 10 at federal and only three at the provincial level.

Despite having such a large number of organisations dedicated to flood control, it was the Pakistan Army that has dominated the scenes of flood response and relief. With a few exceptions, such as Pakistan Meteorological Department and Flood Forecasting Division, the capacity of civil institutions was never enhanced enough for them to take charge of their responsibilities. The flood-focussed long-term disaster planning was marginal, limited to river flooding, centralised and involved federal organisations.

[4] Excluding Punjab Emergency Service, which was operative only in some parts of the Punjab province before the 2005 earthquake.

Table 3.4 Pre-2005 earthquake role of federal ministries and provincial departments in the disaster management cycle

Response	*Relief*	*Preparedness*
National Crisis Management Cell	–	–
Civil Defence Department	Civil Defence Department	Civil Defence Department
–	Emergency Relief Cell	Space and Upper Atmosphere Research Commission
Pakistan Army	Pakistan Army	Pakistan Meteorological Department
Army Aviation	Pakistan Air Force	Flood Forecasting Division
Federal Works Organisation	Federal Works Organisation	Federal Works Organisation
Pakistan Navy	Pakistan Navy	–
–	–	Federal Flood Commission
–	–	Water and Power Development Authority
	–	Dams Safety Council
–	–	The Indus River Commission
–	–	Planning Commission of Pakistan
Provincial Departments[a]		
–	–	Planning and Development
–	Relief Departments	–
Police Department	Police Department	–
Provincial Crisis Management Cells	–	–
Civil Defence Department	Civil Defence Department	Civil Defence Department
Irrigation Department	Irrigation Department	Irrigation Department
–	Health Department	–

(continued)

Table 3.4 (continued)

Response	*Relief*	*Preparedness*
–	Agriculture and Livestock Department	–
–	Food Department	–
–	Communication and Works Department	–
Punjab Emergency Service[b]—Rescue 1122	–	–

[a]The nomenclature of provincial departments varies slightly across provinces, so general terms are used here
[b]Only in the Punjab province before the 2005 earthquake
Source Author

No central agency existed to take full charge of designing disaster policy, and its implementation at federal and provincial levels. Moreover, different responsibilities related to disaster management were fragmented across several institutions. For example, ERC was only responsible for dealing with the post-disaster situation and National Crisis Management Cell (NCMC) was there to ring the danger bells at the time of a disaster.

The Civil Defence Department has a district-level set up and has been made responsible for disaster response since 1993. High- and low-ranking government officials believed that in contrast to its responsibility, it is poorly resourced and woefully ignored by provincial and federal governments alike (Senior Officer, NDMA, Islamabad, 04/10; Instructor, Provincial Civil Defence Department, 05/10). Instead of building its capacity by injecting financial resources and training its staff, another institution, the NCMC, was set up in 1999. In the case of the Punjab province, the largest province population-wise, yet another institution, Punjab Emergency Service 1122, was added to the number of disaster response institutions in 2004. In the absence of an integrated and coherent policy on disasters, the disaster management structure was made more complex by adding new structures without clear boundaries of the mandate over a period. Lately, other provinces have also begun to emulate the model of Punjab Emergency Service 1122. Peshawar, the capital city of the case-study province of this study, launched its rescue service on August 30, 2009 (*The Dawn*, 2009).

3.4.2 *Private Sector*

Having no key institution to engage with a range of stakeholders and relying solely on the Pakistan Army for flood response and relief before 2005, the government did not explore the opportunity of connecting with the private sector for the provision of goods and services in disaster situations. Like a top-down, decision-making process, the disaster management structure was also heavily centralised, hardly leaving any room for other entities beyond the government. The government, the Armed Forces, in this case, began to compete with the private sector by directly entering into the private sector through publicly owned rent-seeking enterprises such as Federal Works Organisation (Siddiqa, 2007). The Frontier Works Organisation is a subsidiary of the Pakistan Army under the Ministry of Defence and an infrastructure building company. It bids in the open market to win construction contracts. Its Chief Executive Officer is always a serving military Lieutenant General. Most often, it was the Frontier Works Organisation that was called on by the Pakistan Army in a disaster situation, and other private sector actors were not encouraged to participate as a matter of government policy. However, on their own accord, private sector actors such as businesspersons, traders and industrialists would regularly provide food, tents, medicines and construction materials to disaster-affected people.

3.4.3 *Civil Society*

Owing to the absence of a bottom-up development approach and inadequate disaster management policy in Pakistan, there was no formal consideration in the government policy to recognise or include civil society in flood control decision-making. Nevertheless, non-governmental organisations and community groups have been working on a self-help basis for counter-flooding measures, such as strengthening the unstable banks of a canal flowing alongside a community.

It would be worthwhile to mention here the scope of the Civil Defence Department. Although ignored and not playing an active role in disaster responses in the current form, it had played an important role in advising ordinary people of safety measures during the 1965 and 1971 wars with India. The Civil Defence Department policies have an elaborate arrangement for the involvement of civil society. This example shows that there is provision for the inclusion of civil society in disaster management policy at the grassroots level, but it has been neglected throughout the country's history.

The actual and potential role of the Civil Defence Department was explored during my fieldwork visit to a district in the Punjab province where I met with a member of the department with more than a decade of service (Instructor, Provincial Civil Defence Department, 05/10). The Civil Defence Department is headed by a District Coordination Officer[5] (DCO) at the district level. The head of district administration, DCO, is the ex officio District Controller Civil Defence. The Civil Defence Department is supposed to register volunteers and train them in different life-saving techniques such as search and rescue, firefighting and first aid. In theory, involvement of civilians occurs when members, preferably apolitical but notable, are taken on as Chief Warden, Additional Chief Warden, Deputy Chief Warden, Divisional Warden, Group Plan and Post Warden in the management of the Civil Defence Department. A "Warden" is referred to as W–willing, A–active, R–resourceful, D–dutiful, E–effective and N–noble, as a way of popularising the idea. These warden positions are honorary and nominated by the DCO. This hierarchy of wardens is to keep a working liaison with other community members who are volunteers for the Civil Defence Department.

However, during the fieldwork, I found thatthe district Civil Defence office I visited was heavily understaffed. This office was supposed to look after the one million people of the district in case of any eventuality with a staff of 12 people (Instructor, Provincial Civil Defence Department, 05/10). This included one DCO/ex officio head of the department, one district officer Civil Defence, one bomb disposal technician, and one bomb disposal expert (both of whom were non-permanent staff and on deputation from the army), three instructors and five secretarial support individuals. There were neither the resources nor the capacity to register, train, and track community volunteers, let alone communicate with civil society members. The three instructors of the department could not even follow up with those who were already registered. The office was not only lacking in human and financial resources but also lacked space and office equipment, like computers, desks and chairs, for the existing staff. The staff members were demoralised and demotivated when asked about

[5] District Coordination Officer was Called Deputy Commissioner before the introduction of the Local Government Ordinance 2001.

the future of their department (Field Journal, May 2010). In consequence, the department was training only school teachers (since they were made to attend training by the district government order), and the school administration was required to provide school buildings as venues for the training.

Even three years after the 2005 earthquake, the same awful condition existed in other provinces, such as Karachi, Sindh's provincial capital. The department was heavily under-resourced, emergency drills were non-existent, the staff were disoriented, and 80 of them posted in four districts of Karachi were not performing any duty due to ignorance of the provincial government to run the department properly (Ali, 2008).

3.5 Pre-2005 Emergency Response System

For flooding, the government had designed a working mechanism for flood control, response, evacuation, relief and preparedness. However, there was no formal and standardised procedure or plan explaining the roles and responsibilities of different stakeholders involved in disaster response, relief and preparedness at any one place. Although a broad understanding existed among different institutions about each other's role, there was also a large grey area. This area would be filled, depending on the mutual rapport and networking of key persons serving in different institutions, since boundaries of organisations could be stretched either way, given the absence of any defined procedure.[6]

During the monsoon season, different organisations including Pakistan Meteorological Department, the Federal Flood Commission, the Flood Forecasting Division, the Pakistan Army and Provincial Relief Department, would yearly exchange information and check the situation. However, this coordination would rarely result in a prompt response. At the time of actual flooding, different organisations such as the Pakistan Army, on the request of the civil administration, Civil Defence Department and Police, along with the affected people's kith and kin, would help in evacuation on an ad hoc basis, lacking unity of command under a single institution. If the Pakistan Army had been called during flooding, it would

[6] Some parts of this section draw on my work during doctoral thesis titled "exploring the role of the mosque in dealing with disasters: A cases study of the 2005 earthquake in Pakistan" at Massey University, New Zealand available at http://hdl.handle.net/10179/4080.

lead the response operation as it had trained personnel and resources. Civil departments such as Civil Defence and Police would aid the army. After the response phase, the civil departments would follow up with the affected population providing relief goods with the help of the army. The relief would be dealt with chiefly by Provincial Relief Departments. The Pakistan Navy and Air Force would also run under the command of the army if requisitioned by the civil administration. After a few days, once the flood had subsided, people would be left to themselves to settle and recover on a self-help basis, with many government promises that were rarely fulfilled (Senior Officer, NDMA, Islamabad, 04/10). The 2005 UNISDR country assessment report noted that there were "no long-term, inclusive and coherent institutional arrangements to address disaster issues with a long-term vision" (Flood Forecasting Division, 2005). It was not until the 2005 earthquake that the government set out a clear rehabilitation programme for the affected population and set up an authority to implement it.

3.6 Influence of Local Government Ordinance 2001

Local Government Ordinance 2001 was important legislation, though not causally related to disaster management, providing a noteworthy diversion from the earlier legacy of top-down and centralised decision-making to a broad-based and grassroots type of governance structure. It influenced future disaster management structure and policy. The Ordinance was introduced in all the four provinces and the first local government elections were held in August 2001. Under this local government system, the district administration was put under the elected public representative. Three tiers of the administration having financial and administrative powers were introduced: district, tehsil and union council. The provinces were made up of districts. Among the districts, there were urban districts and rural districts. Urban districts were divided into municipalities and towns; rural districts were divided into tehsils for administrative purpose. Tehsils were further sub-divided in Union Councils, the lowest tiers of the state structure.[7]

[7] Some parts of this section draw on my work during doctoral thesis titled "exploring the role of the mosque in dealing with disasters: A cases study of the 2005 earthquake in

Different safety functions, such as flood control protection, stormwater drainage and civil defence planning including rapid response contingency plans, were to be designed and implemented with the active participation of the concerned communities and their representatives at district-, tehsil- and union council levels. It was proposed that several governance structures that involved substantial participation from civil society, such as the Public Safety Commission and the Community Citizen Boards, were to monitor the working of the district bureaucracy. During the time of the 2005 earthquake and later, during the conception and initiation of the new NDRMF in 2007, the country was under a military ruler (General Pervez Musharraf) and the local government system had the full support of the government. Therefore, the layout, design and suggested institutional structure of the NDRMF were connected fully to the local government system.

Former President, Pervez Musharraf, promulgated a presidential ordinance, the National Disaster Management Ordinance, on December 23, 2006, with an elaborate role for local governments in line with the spirit of Local Government Ordinance 2001. This ordinance supported necessary institutional arrangements, policy guidance and a reference point for the future disaster management structure in Pakistan. The ordinance was later made an Act of Majlis-e-Shoora (Parliament) in 2010 and now called National Disaster Management Act, 2010. The Act keeps clear roles for district disaster management authorities assuming fully functional local governments.

However, the phenomenon of a representative government disowning local government reforms introduced by a non-representative government is centuries old in pre- (the first municipal Madras corporation in 1688, the Karachi municipality in 1852, and the 1912 Panchayat Act) and post-independence (the Basic Democracies Ordinance 1952 and Local Government Ordinances 1972) Pakistan (Ali Cheema et al., 2005).

3.7 Key Challenges of the Pre-2005 Arrangements

The Quetta earthquake had occurred only 12 years before the creation of Pakistan but seemed to have provided no lessons for approaches to disasters in the country. It is clear from more than five decades of development planning in Pakistan, shown through the analysis of 10 5-year

Pakistan" at Massey University, New Zealand available at http://hdl.handle.net/10179/4080.

plans (1955–2010), that flooding has been the most recurring hazard with which the government has dealt. The pre-2005 earthquake disaster policy and the ensuing management structures targeted the reduction of river-flood losses and invested in structural measures such as the construction of dams, barrages, embankments and a few non-structural measures such as the collection of real-time flood data, a flood early warning system and coordination for evacuation of affected communities. The issue of reduction of losses from torrential rains and flash flooding, however, was not addressed.[8]

Adhering to a contingency and reactionary approach towards disasters, the government kept on adding new structures to address the emerging needs of the time, hoping to reach efficiency and improvement in the process. For example, the Federal Flood Commission was set up in 1977 when the provincial flood management organisations could not deal with severe floods in 1973 and 1976. However, the addition of new actors in the flood management chain only marginally improved the flood control system and protection of life and property. It certainly resulted in further distribution of responsibilities, which meant that no single institution could be held responsible for a failure. Similarly, no serious endeavour was made to enhance the capacity of the related provincial line departments such as the Civil Defence Department.

Since its establishment in 1947, Pakistan has remained under the rule of military[9] generals for half its life. The capacity of civil institutions could not be enhanced owing to frequent disruption of the political process (A Cheema et al., 2006). To fill the gap, the Pakistan Army was invited to capture the centre stage of flood response and relief. The army has a trained workforce, resources and organisational skills to respond to a disaster situation. However, the army has not undertaken long-term DRR strategies and disaster management policies. A few civil institutions, such as the Pakistan Meteorological Department, Space and Under Atmosphere Research Centre, and the Flood Forecasting Division were upgraded, and their technical capacity was developed in terms of human

[8] Some parts of this section draw on my work during doctoral thesis titled "exploring the role of the mosque in dealing with disasters: A cases study of the 2005 earthquake in Pakistan" at Massey University, New Zealand available at http://hdl.handle.net/10179/4080.

[9] So far, periods of military rule include 1958–1971, 1977–1988 and 1999–2007.

and technical resources after the 1992 floods. This improvement in technical capacity (with the assistance of the Asian Development Bank) has contributed to an improvement in flood forecasting mechanisms.

In general, national armies play a role in disaster response, domestically and internationally, in many countries, including developed countries, such as the USA (Harrison, 1992; Thompson, 2010). Nevertheless, this does not absolve civil authorities of their responsibilities for disaster preparedness, as has been the case in Pakistan. The army's strong position in the country's overall growth process pushed out the participation of civil institutions. This over-reliance on the army in the disaster response, coupled with neglect of long-term disaster preparedness strategies in the country, had severe implications (A. R. Cheema et al., 2016). The consequences of this lack of preparation became clearer at the time the 2005 earthquake occurred, and the nation paid a heavy price in terms of human and economic losses.

It is surprising to note that the government included the multi-hazard approach to disasters in the Medium-Term Development Framework just 4 months before the 2005 October earthquake. Similarly, another report by the ISDR, released in January 2005, showed that Pakistan's disaster management structure was out-of-date, narrowly focussed on floods and lacked long-term cohesive institutional arrangements to address disaster issues. However, the 2005 earthquake did not allow the government to prepare, and it caught people by surprise.

From the analysis of developmental plans of the country over the last 55 years, it appears that there was inertia[10] (resistance to change) in the disaster management structure of the country. There could be three key reasons for this inertia. Firstly, Pakistan had not faced a high-scale calamity like that of the 2005 earthquake that could have become a strong reference point to sensitise the pattern of future disaster policymaking. Secondly, the country struggled to meet the pressing needs of its growing population, such as poverty, health and education, thus it was difficult to free up resources for future disaster planning. Thirdly, the existing disaster management institutions of the country could not suggest or implement the required changes in disaster infrastructure and policy due to capacity

[10] This inertia in an institutional structure has been of interest to academic scholarship and is referred to as path dependence (Imran, 2010; Jacob, 2001).

challenges. When the change in disaster management policy and infrastructure was introduced under the influence of a donor (UNDP in this case), it was not followed by actions until the 2005 earthquake occurred.

Like the centralised decision-making of pre-2005 policy, the disaster management structure became heavy at the top, allowing only a marginal role, if any, for the private sector and civil society. Instead of encouraging broader participation of other private sector players as a matter of policy and practice, the government relied on a single army-owned government agency, the Frontier Works Organisation. Likewise, the disaster policy decision-making was thought to be too serious a business to be taken to civil society, therefore, a technocratic mindset and approach prevailed. The affected communities were coordinated to the extent of dissemination of advanced flood warning and mosques were engaged for announcements only. However, the involvement of communities in disaster management such as local disaster preparedness plans was non-existent.

Overall, the flood-centric policy framework and fragmented responsibilities of different disaster management institutions show the lack of an effective institutional disaster management structure for the reduction of disaster losses in Pakistan, particularly at the local level. The disaster policy relied only on reaching the short-term aims of immediate response and relief while ignoring the long-term goal of risk reduction through paying attention to the improvement of disaster management in the country. The Local Government Ordinance, introduced in 2001, was an attempt to devolve power to the grassroots level and to involve the poor and the marginalised in the process of making decisions that affect their lives, such as the formulation of disaster management contingency plans. This piece of legislation has had a major effect on the structure and suggested institutions in the NDRMF, Disaster Management Ordinance 2006 and finally Disaster Management Act 2010.

3.8 Interim Disaster Management Policy and Structure

It was at once realised by the federal government that the magnitude of loss from the 2005 earthquake was enormous, and the existing institutional structure had neither the capacity nor the resources to handle it. Therefore, a new institution, namely, the Federal Relief Commission

(FRC) was set up to deal with this extraordinary situation in the aftermath of the destruction caused by the earthquake.[11]

3.8.1 Federal Relief Commission

It has been noted in the NDRMF that the inability of Pakistan's emergency response system was exposed after the 2005 earthquake. FRC was set up on October 11, 2005, 2 days after the earthquake, to coordinate the huge response and relief operation needed in the earthquake-affected areas. A senior army officer of the rank of Lieutenant General was appointed as the first relief commissioner. There was an immense response from within the country as volunteers loaded with relief items looked to reach the earthquake-affected people.

The gigantic task of the commission was to organise and structure the response and relief operation in partnership with federal and provincial departments, civil society organisations and the international community bringing relief aid and human resources, such as response workers and paramedics. The relief was formally ended by the government on March 31, 2006, and FRC was dissolved. The establishment of the FRC was a stop-gap arrangement and the government felt the need for another organisation that could take the job beyond relief. The unprecedented task of reconstruction and rehabilitation in Pakistan's history was therefore handed to another authority already created by the government on October 24, 2005.

3.8.2 Earthquake Reconstruction and Rehabilitation Authority

The Earthquake Reconstruction and Rehabilitation Authority (ERRA) was established as a statutory body through a presidential ordinance on October 24, 2005, mainly to take up the enormous task of rebuilding an earthquake-affected region spread over 30,000 square kilometres (Earthquake Reconstruction & Rehabilitation Authority, 2005). ERRA took over from FRC. Initially, the core group of ERRA was made up of civil bureaucrats, armed forces personnel and international consultants. Except

[11] Some parts of this section draw on my work during doctoral thesis titled "exploring the role of the mosque in dealing with disasters: A cases study of the 2005 earthquake in Pakistan" at Massey University, New Zealand available at http://hdl.handle.net/10179/4080.

for the chairman, who is a civilian having more of an advisory role, the operational position of deputy chairman, second in command after the chairman, and many other key positions in the organisation have, since its establishment, been occupied by army officers, although it is a civilian institution on paper.

ERRA was a project-based organisation meant only for the reconstruction and rehabilitation of the nine earthquake affected districts of Khyber Pakhtunkhwa and Azad Jammu and Kashmir. Its aim was to streamline all activities related to post-disaster damage assessment, reconstruction and rehabilitation under one roof to promote the pace of development in the earthquake-affected areas. Two other authorities, the Provincial Reconstruction and Rehabilitation Authority and the State Reconstruction and Rehabilitation Authority, were established in the North West Frontier Province (now renamed as Khyber Pakhtunkhwa) and Azad Jammu and Kashmir (AJ&K), respectively. These authorities were to undertake reconstruction and rehabilitation activities in their areas while ERRA was to provide funds and set guidelines. At that time, these authorities were set up to complete the project of reconstruction and rehabilitation in the nine earthquake-affected districts in 3 years. The evaluation of ERRA's performance is beyond the scope of this book. Prime Minister's Cabinet has approved the subsuming of ERRA in the NDMA as part of the institutional mainstreaming endeavour (*The Express Tribune*, 2019). However, its policies are referred to where they are linked to the role of the mosque in disaster management.

3.9 Post-2005 Disaster Management Policy

A new institutional policy framework for disaster management was proposed in the aftermath of the 2005 earthquake. Former President, Pervez Musharraf, promulgated a presidential ordinance, the National Disaster Management Ordinance, on December 23, 2006. This ordinance provided for necessary institutional arrangements, policy guidance and a reference point for the future disaster management structure in Pakistan. The ordinance was later made an Act of Majlis-e-Shoora (Parliament) in 2010 and now called National Disaster Management Act, 2010. In the same year, Disaster Management was made a provincial subject through 18th amendment to the Constitution of Pakistan. In the aftermath of 18th constitutional amendment, all the provinces are responsible to manage disaster in their own capacity.

The NDRMF was prepared by National Disaster Management Authority (NDMA) with the technical assistance of the UNDP Pakistan in March 2007. NDMA is the apex federal disaster management body of the country. This organisation is required to prepare a National Disaster Response Plan under section 10 of the National Disaster Management Act and then seek its approval from the National Disaster Management Commission (NDMC). Provincial, district, municipal and town level disaster management authorities have been identified and their responsibilities have been shown in the Act. As part of the institutional mainstreaming, the government merged the Emergency Relief Cell in NDMA in 2015. This Cell was set up in 1971 at the federal level to deal with the emergency in the aftermath of the cyclone in East Pakistan. Its job was to stockpile goods and relief items and coordinate with provincial relief departments. It operated as emergency control room to liaise with other departments. Now this stockpiling and coordination role has been taken over by the NDMA.

Before the Act, the NDRMF highlighted and elaborated upon the roles and responsibilities of all the stakeholders involved in disaster management. It emphasised that all stakeholders had to help in initiating the three basic activities in case of a disaster. These activities were to conduct damage and loss assessments after a disaster, coordinate emergency response, and participate and organise recovery and rehabilitation initiatives in line with duties of a department.

Figure 3.1 shows some of the key stakeholders involved in disaster management including all provincial and state disaster management authorities, Government Departments, United Nations and civil society organisations namely Pakistan Humanitarian Forum (PHF) and National Humanitarian Network (NHN) and media. Armed Forces are an important stakeholder and are engaged by NDMA when needed.

3.10 Post-2005 Disaster Management Structure

Under the National Disaster Management Act (NDMA) 2010, elaborate decision-making and implementation bodies were suggested for three distinct levels of administration: federal, provincial and district. The Act has followed the layout of the Local Government Ordinance 2001 mentioned in the preceding sections of this chapter in terms of devolution of disaster management responsibilities. Pakistan has a federal structure

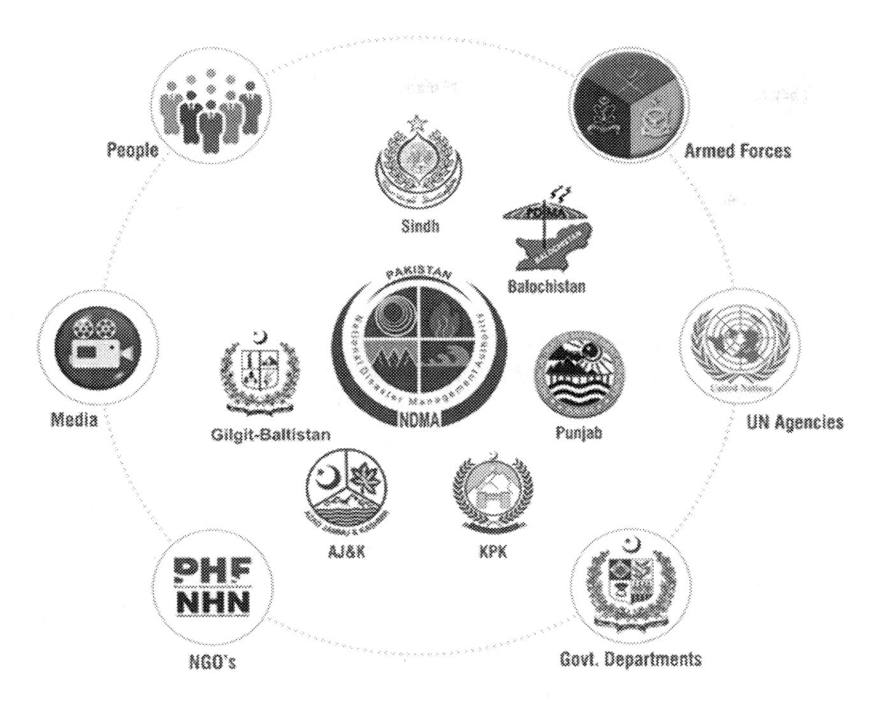

Fig. 3.1 Stakeholders involved in Disaster Management with the central role of NDMA (Armed Forces = Pakistan Army, Pakistan Navy and Pakistan Air Force; UN agencies = All United Agencies working in Pakistan such as United Nations Development Programme (UNDP), Food and Agriculture Organisation (FAO) and World Food Programme (WFP); Provincial Disaster Management Authorities (PDMAs) = Balochistan, Khyber Pakhtunkhwa, Punjab and Sindh; Azad Jammu & Kashmir (AJ&K) = State Disaster Management Authority; Gilgit-Baltistan (GB) = GB Disaster Management Authority; PHF = Pakistan Humanitarian Foundation; NHN = National Humanitarian Network. *Source* Author, amended from NDMA [2021])

of government with a bicameral legislature. It has four[12] federating units

[12] Gilgit-Baltistan (GB) Assembly has passed a unanimous resolution in March 2021 demanding of the Government of Pakistan to grant the "Provisional Provincial Status" to Gilgit-Baltistan region till the Kashmir issue is settled through plebiscite according to the UN Security Council resolution. Currently, GB has a Council consisting 24 members

called provinces—Punjab, Sindh, Khyber Pakhtunkhwa and Balochistan and two administrative units namely, Azad Jammu and Kashmir (AJ&K) and Gilgit Baltistan (GB).

3.10.1 Government: Federal Level

A country-level disaster management commission, named National Disaster Management Commission (NDMC) and National, has been set up at the federal level. The NDMC is headed by the Prime Minister with Chief Ministers of the four provinces and also Ministers of Defence, Health, Foreign Affairs, Social Welfare and Special Education, Communications, Finance, Interior, Leaders of Opposition in Senate and National Assembly, among others. The Commission has been dormant for most of the time and could held only five meetings so far, the last one on March 28, 2018 until 2020, 10 years since the Act was passed in 2010.

The Act provides for the establishment of a federal disaster management authority, NDMA, to function as the operational arm of this commission. NDMA serves as the secretariat to implement the decisions of the commission and to implement, coordinate and monitor the implementation of related national policies. NDMA is also mandated to prepare a National Disaster Response Plan. NDMA is also responsible for providing necessary technical assistance to the Provincial Governments and the Provincial Authorities for preparing their disaster management plans following the guidelines laid down by the National Commission. The authority is responsible for coordinating reaction in the event of a potentially dangerous crisis situation or disaster. The National Disaster Response Plan 2019 sets a vision for a better prepared Pakistan by investing in disaster preparedness, coordination mechanisms and effective and efficient disaster response for minimising losses to human life, livelihoods, infrastructural and environment.

3.10.2 Government: Provincial Level

In the new disaster management institutional structure, each province has a Provincial Disaster Management Commission (PDMC) and the

directly elected by the people while the prime minister of Pakistan as its Chairman. The chief minister and the governor are also in place but the region does not the status of a province (Shaheedi, 2021).

two administrative units of Azad Jammu & Kashmir (AJ&K) and Gilgit-Baltistan (GB) have State Disaster Management Authority (SDMA) and Gilgit-Baltistan Disaster Management Authority (GBDMA), respectively. Like NDMC, PDMC is headed by the respective Chief Minister and Ministers of Finance, Revenue, Law, Leader of Opposition in the provincial assembly among others. The Provincial Disaster Management Authority (PDMA) functions as an operational arm of the PDMC. The federal government has notified PDMCs and PDMAs for the four provinces of Pakistan.

3.10.3 *Government: District, Tehsil and Union Council Levels*

Under PDMAs, District and Municipal Disaster Management Authorities have been proposed. Municipal/Tehsil and Town disaster management Authorities would function under District Disaster Management Authorities (DDMAs) in urban and rural areas, as shown in Fig. 3.2. National Disaster Management Act 2010 is the overarching national legislation. Under the Act, the National Disaster Management Commission (NDMC) provides national oversight in policy formation while Provincial Disaster Management Commissions (PDMCs) operate at each of the four provinces of Pakistan to enable the achievement of national policies.

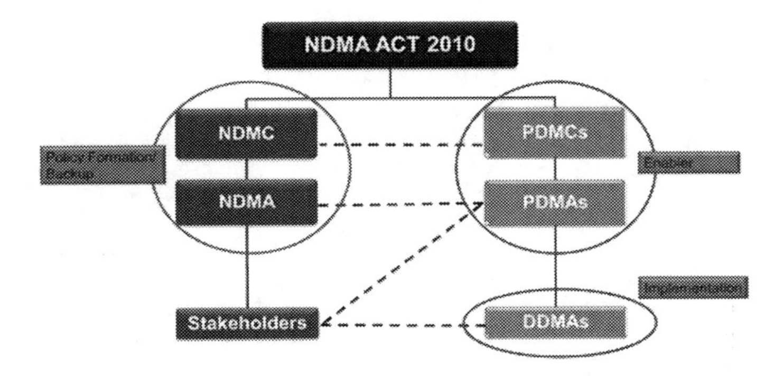

Fig. 3.2 Disaster Management Structure of Pakistan (NDMC = National Disaster Management Commission; NDMA = National Disaster Management Authority; PDMCs = Provincial Disaster Management Commissions; PDMAs = Provincial Disaster Management Authorities; DDMAs = District Disaster Management Authorities. *Source* Author amended from NDMA [2021])

National Disaster Management Authority (NDMA) functions under the guidance of NDMC and coordinates with Provincial Disaster Management Authorities (PDMAs) and with other stakeholders. PDMAs operate at the provincial level and coordinate with District Disaster Management Authorities (DDMAs) to ensure effective implementation of disaster management plans and policies.

Further to District and Municipal Disaster Management authorities, the Disaster Response Plan mandates Tehsil and Town Disaster Management Authorities to implement national, provincial, district and local disaster mitigation guidelines and assist in carrying out disaster response and relief activities. The plan envisages that a number of government departments at grassroots level, such as agriculture through its extension workers, police, education, health and revenue, will come into direct contact with communities at this level of administration to carry out their different roles in different phases of disaster management.

Tehsil comprise union councils, the lowest tier of administration. The NDRMF envisioned an initiative-taking role for these councils. They were expected to stand for their people fully through their elected representatives and therefore to advocate for their needs to higher disaster management authorities to seek resources. They must also seek allocation of resources from higher authorities for vulnerability reduction activities including spurs for flood control and rainwater harvesting structures for drought mitigation. The NDRMF pointed where "capable union councils may develop local policies and guidelines for vulnerability reduction" (National Disaster Management Authority, 2007).

The DDMA has the responsibility to formulate district disaster plans, its implementation, coordination with the key stakeholders, and give directions to other departments and authorities to take necessary measures for prevention and mitigation. At the lower tier of institutional structure, the elected town heads, sub-district heads and union council heads are to serve as the frontline local leaders to respond to disasters.

Later, the Act and 2019 Disaster Response Plan also place the immediate responsibility of the safety, shelter and food and nutrition on DDMAs. However, the Act and the earlier NDRMF assume functioning local government system.

The Constitution of Pakistan requires meaningful devolution of political, administrative and financial powers to local governments under Article 37 and Article 140-A. However, in Pakistan, local governments have been struggling for meaningful political, administrative and financial devolution particularly in the aftermath of the 18th Amendment to the Constitution of Pakistan in 2010. Though this amendment allowed greater autonomy to the provinces, the provinces have held control, marginalised local governments and kept authority to direct and oversee them. Though different amendments have been made to the local government acts by the respective provinces, local governments, both in urban and rural areas, still are deprived of political, administrative and financial autonomy to effectively discharge their functions. The local governments in each of the provinces differ in terms of their structure such as tenure, election process and fiscal powers.

A range of functions such as disaster management, public health, primary education, water supply, sanitation, town planning and building control have been devolved to different tiers in the local government system by the provinces. However, none of the local governments can levy a tax or incur debt without approval of the respective provincial government as per the prevailing laws governing local governments in the four provinces. Effective management of basic services requires closer cooperation between local authorities and other levels of government; improved vertical and horizontal coordination between local, regional, national and international institutions is necessary. Effective multi-level governance requires institutional and legal frameworks that clearly define the roles and responsibilities of all levels of government, guided by the principle of subsidiarity.

The local governments formed across the country in 2014 and 2015 have completed their tenures. New elections have not been held yet until March 2021. Historically, provincial governments tend to postpone local government elections and lately COVID-19 has provided a pretext. In lower middle countries, such as Pakistan, gaps between required investment and current resources in basic services are the widest in rural areas.

Typically, at the district and sub-district levels, where emergency response should have been the strongest in terms of human, administrative, financial and technical resources, still is ignored and marginalised. The Deputy Commissioner, administrative head of the district, has an ex-officio charge of the head of District Disaster Management Authority. The Deputy Commissioner, owing to a range of other responsibilities and frequent turn over, is unable to pay due attention to disaster preparedness and risk reduction measures. It's only at the time of emergency that district administration moves to evacuate and respond to an emergency. The understanding of DRR remains extremely limited among officials of line departments and local political leadership indicating the weakest link where it should be the strongest (I. Shah et al., 2020).

Despite elaborate institutional disaster management structure at the top (federal and provincial levels), the frontline government departments who have primary responsibility of saving lives, assets and mitigating disaster losses are still marginalised and ignored.

3.11 Post-2005 Emergency Response System

Learning from the 2005 earthquake experience when Federal Relief Commissioner had to be established to deal with the catastrophic situation, a new emergency response system was devised in the NDRMF and also explained and updated with detailed responsibilities in the National Disaster Response Plans of 2010 and 2019, respectively. Under this system, Emergency Operation Centres at national, provincial and district levels have been set up and named as National Emergency Operation Centre, Provincial Emergency Operation Centre and District Emergency Operation Centre, respectively. The National Emergency Operation Centre, Provincial Emergency Operation Centres and District Emergency Operation Centres are under the supervision of NDMA, respective PDMAs and DDMAs, respectively.[13]

The *modus operandi* of the working of Emergency Operation Centres has been explained in the National Disaster Response Plan released in 2010, the first ever response plan in the history of Pakistan, and then in 2019 (National Disaster Management Authority, 2019).

According to the latest available National Disaster Response Plan of 2019, the District Emergency Operation Centre will operate in the case of a local-level emergency and the concerned DDMA will mobilise its resources to deal with the situation. The District Emergency Operation Centre will coordinate with all relevant government agencies and civil society organisations for response, relief, and rehabilitation of the affected population. In the case of a disaster exceeding local capacity, the DDMA

[13] Some parts of this section draw on my work during doctoral thesis titled "exploring the role of the mosque in dealing with disasters: A cases study of the 2005 earthquake in Pakistan" at Massey University, New Zealand available at http://hdl.handle.net/10179/4080.

will request assistance from the concerned PDMA. The Provincial Emergency Operation Centre will be activated and will help organise resources for response, relief and rehabilitation of the affected population. In the case of a disaster exceeding the provincial capacity, the PDMA will request NDMA for assistance and the National Emergency Operation Centre will be activated. At this stage, focal persons of the federal ministries and departments are called at the centre to allow fast coordination and implementation of emergency operations. The National Emergency Operation Centre will then help organise response, relief and rehabilitation of the affected population. In case of a national emergency, the Prime Minister would announce a national emergency and request international assistance if so required.

3.12 Role of Other Government Institutions in the Post-2005 Disaster Management Policy and Structure

A multi-hazard approach has been adopted and the number of government institutions, organisations and statutory bodies involved in disaster response, relief, recovery and preparedness has been increased substantially compared with the pre-2005 earthquake situation. Table 3.5 shows the roles of 40 federal ministries, departments and authorities and in disaster management.

In line with the Disaster Management Act 2010 and guidelines of the National Disaster Response Plans, provincial governments have also set up elaborate disaster management structures to examine the vulnerability of different parts of the province to different disasters and specify prevention or mitigation measures. Though members of such decisions making bodies might vary from province to province to suit circumstances of any province and administrative unit, Table 3.6 enlists members along with departments of the Punjab Disaster Management. These departments are responsible for policy formulation, coordination and implementation and lay down guidelines for preparation of Disaster Management Plan by provincial departments and district authorities; to evaluate preparedness at all governmental and non-governmental levels to respond to disasters and to enhance preparedness and advise them regarding financial matters in relation to disasters management. Similar arrangements are in other provinces, AJ&K and GB. The Khyber Pakhtunkhwa province,

Table 3.5 Federal ministries, departments and authorities involved in disaster management

Federal Ministries	*Federal Departments and Authorities*
Ministry of Defence	Pakistan Armed Forces
Ministry of Interior	Pakistan Meteorological Department (PMD)
Ministry of Foreign Affairs	Civil Aviation Authority (CAA)
Ministry of Communications	Federal Flood Commission (FFD)
Ministry of Climate Change	Geological Survey of Pakistan (GSP)
Ministry of Finance, Revenue and Economic Affairs	Survey of Pakistan
Ministry of Planning, Development and Special Initiatives	Indus River System Authority (IRSA)
Ministry of National Food Security and Research	National Database and Registration Authority (NADRA)
Ministry of Housing and Works	National Disaster Risk Management Fund (NDRMF)
Ministry of Human Rights	National Highway Authority (NHA)
Ministry of Industries and Production	National Logistic Cell (NLC)
Ministry of Information, Broadcasting, National History and Literary Heritage	Pakistan Coast Guard
Ministry of Information Technology & Telecommunication	Pakistan Commissioner for Indus Waters (PCIW)
Ministry of Law and Justice	Pakistan Housing Authority
Ministry of Energy	Pakistan Public Works Department
Ministry of Maritime Affairs	Pakistan Railways
Ministry of Railways	Police Service of Pakistan
Ministry of Health Services, Regulation and Coordination	Press Information Department (PID)
Ministry of Science and Technology	Space and Upper Atmosphere Research Commission (SUPARCO)
Ministry of Water Resources	Water and Power Development Authority (WAPDA)
Prime Minister Office	
	National Command and Operation Centre (NCOC)

Source Author. Amended from National Disaster Response Plan, NDMA (2019)

the provincial government made an amendment in the NDM Act 2010 known as National Disaster Management Khyber Pakhtunkhwa Amendment Act 2012. Under the act, the provincial government changed the name and structure of DDMA. The DDMA has was changed to a District Disaster Management Unit (DDMU) in district administration (I. Shah et al., 2020).

Table 3.6 Members of the Provincial (Punjab) officials and departments responsible for formulation of disaster management policy

Provincial Officials and Departments
Relief Commissioner (Chairperson)
Secretary Home
Secretary Law
Secretary Finance
Secretary Health
Secretary Agriculture
Secretary Irrigation
Secretary Housing and Public Health Engineering Department
Secretary Local Government
Secretary Social Welfare
Secretary Livestock
Secretary Information
Additional Inspector General of Police (Operations)
Director General PDMA (Secretary)
Director General Punjab Emergency Service
Director General Civil Defence
Any other members to be co-opted by the Chairperson

Source Author based on Government of the Punjab (2019)

3.13 Key Challenges of the Post-2005 Arrangements

The NDRMF enlisted a number of new government organisations with new roles in the disaster management cycle have been included in the national disaster management structure. This has led to a beginning of a paradigm shift in the disaster management policy and practice in Pakistan.

However, analysis in this section shows that the addition of new actors increased the complexity of the disaster management structure and created new institutional challenges without deciding the role of the existing ones. Ad hocery and short-termism coupled with adherence to the old pattern of erecting new structures without evaluating, upgrading or dismantling the former have again emerged.

To add further complexity, friction and power conflict, the National Oversight Disaster Management Council was established in the aftermath of the 2010 flooding superseding the NDMC created in 2006 for the same purpose of overseeing disaster preparedness, relief and rehabilitation operations of NDMA.

Parallel government set ups are competing for resources, creating an environment of conflict to justify their existence. This power struggle and conflict of interest is counterproductive for the cause of disaster management in the country. A living example of these counterproductive endeavours is visible between the institutional struggle between NDMA and the Ministry of Climate Change (MoCC). The MoCC has a broader mandate of dealing with climate change through national policy, plans strategies and programmes with regard to disaster management including environmental protection, preservation, pollution, ecology, forestry, wildlife, biodiversity, climate change and desertification. However, NDMA functions independently of the MoCC directly under the supervision of the Prime Minister's Office.

These organisations frame their policies and strategies according to their own interests and priorities. They are not necessarily in coordination with other relevant federal ministries and provincial departments. These policies are designed by high-level officials in consultation with donors, and not necessarily through participatory processes which might have aligned them with ground realities. Donor funded consultants prepare policies and they rarely consult district-level line departments or local communities who have lived with climate change and faced disasters for decades. This institutional disconnect multiplies in terms of forestalling synergies in mainstreaming climate change adaptation and DRR across economic, social and environmental streams of development. For policy coherence and to realise the goal of a climate change and disaster resilient Pakistan, sustainable development, NDMA should have been made to work under the MoCC.

In the COVID-19 context, another organisation named National Command and Operation Centre (NCOC) has been set up, again headed by a serving military officer along with several other officer in lower ranks. Given the broad scope of NDMA, which includes health emergencies, the COVID-19 reaction should have been implemented via NDMA. Pakistan is rapidly urbanising with rising population growth and rural to urban migration. Another key challenge to effective DRR is ineffective implementation of building byelaws in urban and rural areas. This task has been further complicated due to the 18th Amendment where PDMAs have weak capacity to instil climate smart urban planning and compliance of building byelaws. The institutional disconnect is more pronounced among various line departments in provinces where integrated disaster and climate smart land use planning in urban and rural contexts is still fragile. Due to this institutional disconnect at the federal level between NDMA and MoCC, NDMA remains more focussed on disaster response and relief than on mainstreaming DRR and climate change mitigation and adaptions programmes throughout the country and in collaboration with provinces and administrative regions.

Yet another challenge is acute shortage of disaster management professionals in the disaster management organisations across Pakistan. Majority of the NDMA staff and similarly in PDMAs, does not have professional qualification and borrowed from other departments. The disaster management authorities lack service structure and operate on ad hoc basis. To make the matters worse, these staff from other departments avail multiple capacity building opportunities in terms of professional training within and outside the country and finally are repatriated to their parent departments. Consequently, disaster management authorities severely lack skilled human capital, knowledge and experience necessary to address such complex issues.

A case in point is NDMA that has been often headed by a serving senior military officer along with other military officers in other roles. This clearly shows that disaster management has not received due attention as a specialised field to have professionally qualified and skilled human resource. Military forces aid civilian administration in disaster response. However, this over-reliance on the army in the ongoing disaster management structure during normal times has severe implications. NDMA has been at loss in building effective relationships with other federal ministries, PDMAs, private sector and civil society role-players. With its reliance on donors and borrowed labour to carry out its technical role in disaster management throughout the country, the federal authority has been unable to become a learning organisation. Ad hocery persists though lesser than before, even after the full overhaul of the disaster management policy and structure in the aftermath of the 2005 earthquake and, as such, hinders the growth and development of competent disaster management organisations.

On paper, the NDRMF and 2019 Disaster Response Plan (the latest available) emphasise the importance of community-based disaster (National Disaster Management Authority, 2007, 2019). It highlights the role of local organisations and social activist groups in this regard. It maintains that district and tehsil disaster management authorities would promote and strengthen the existing community-based organisations. In case there are no community-based organisations, new social community groups would be formed to promote disaster management activities. Coupled with the formation of new community-based organisations is training in first aid, early warning systems, search and rescue, evacuation and firefighting. The framework envisages that such community-based organisations would be linked with local public service providers such as veterinarian facilities, banks, post offices and agriculture services. Nevertheless, the framework is silent about any formal structure to link community-based organisations with tehsil or district authorities.

Since there was a huge response from civil society in the aftermath of the earthquake when people offered material and personal support, a new organisation named National Volunteer Movement (NVM) was formed on November 1, 2005 (Government of Pakistan, 2011). This came under the Ministry of Youth Affairs with its office in Islamabad. The NVM aimed to channel and mainstream the large number of volunteers from within and outside the country, national and international associations converged on the earthquake-affected areas by linking them with civil society organisations during relief and rehabilitation activities. The NVM remained active during the military government of President Pervez Musharraf, and the NDRMF has called for its strengthening, along with other weaker organisations such as Civil Defence Department.

According to the Sendai framework, the ultimate promise of safer communities lies at the local level. The enabling, guiding and coordinating role of national and federal State Governments stays essential. However, it is necessary to empower local authorities and local communities to reduce disaster risk, including through resources, incentives and decision-making responsibilities (UNISDR, 2015c).

However, due to the lack of attention and allocation of resources to disaster management at national and provincial levels by federal and provincial governments, respectively, local-level disaster management bodies such as DDMAs, Municipal Disaster Management Authorities and Tehsil Disaster Management Authorities remain too handicapped to be able to perform their functions. For example, I visited the first office of the DDMA Mansehra during my second fieldwork period. The salary of the project officer was being paid by GTZ. He had only three people working for him and they too were borrowed from other departments. The DDMA office made up two temporary containers. The project officer stated there was lack of ownership of the DDMA by district and provincial government. He was facing a dearth of human and financial resources to run everyday office correspondence because the policy decision of having DDMAs was not followed by resource transfer. Despite a few successes, like the establishment of search and rescue teams, there were serious concerns about the sustainability of disaster management projects at community level. The functioning of the first DDMA office relied on contingency plans run on a day-to-day basis. It may be true on paper that DDMA was to take care of the Community-based Disaster Risk Management (CBDRM) programme after ERRA but it did not seem to happen in the actual circumstances faced by DDMA.

The CBDRM programme was implemented at Union Council level, the lowest tier of administration, and it was claimed that it had raised the capacity of communities who would stand on their own in a future disaster (Earthquake Reconstruction and Rehabilitation Authority, 2010, 2011). Despite the project's implementation at the Union Council level, the communities under study considered it insufficient to meet the target of proposed capacity enhancement for disaster safety. During my fieldwork in 2010, I joined an ongoing CBDRM programme in one Union Council. This Union Council had a population of 30,000, with 10 major villages located in an isolated, rugged and mountainous landscape in Khyber Pakhtunkhwa. In the programme, there were 20 men and 12 women. These 32 participants included six representatives of different government departments, such as health, revenue and education. Therefore, less than one percent of the people participated at one time in training under the CBDRM programme. The villagers raised strong concern about the insufficient coverage of this training and suggested the programme should be extended to the village level to be significant. Even during the training, certain areas of training, such as climbing down from a hill using a seat harness could not be undertaken because of the low quality of ropes available to the master trainers. The community was also wary about the supply of equipment after the training. Similarly, the master trainers commented on the leakages in funds, which affected their mobility while accessing remote sites. Overall, DDMAs were constrained by lack of human and financial resources; it did not seem that CBDRM would be sustained even in its limited capacity.

The two master trainers interviewed for this study, who I joined in the actual first aid training sessions, explained that it was important to win the support of the imam. The imam was an opinion maker and his views and religious interpretations influenced a community's worldviews.

The master trainers explained how they would proceed (ERRA Master Trainers, CBDRM Project, 4/10):

> In the first meeting with a community including the imam of the village mosque, we would begin with the story of Prophet Noah's Ark mentioned in the Holy Quran. We would explain that Noah was the Prophet of Allah and Allah could have saved His Prophet without asking him to prepare for the flood. But the purpose of asking Noah for preparation and mentioning this story in the Holy Quran is to make us aware that we ought to prepare for disasters.

The engagement with the imam was important for ensuring that the CBDRM initiative was inclusive, particularly of women. The CBDRM programme guidelines also suggested the inclusion of the religious leader in the Union Council Disaster Management Committee. A regional manager and a social mobiliser of the UN-Habitat working to promote safe house building techniques among village people had a similar experience. They explained they had a special orientation programme for imam*s* and then invited them to deliver a speech before house building training workshops (Regional Manager and Social Mobiliser, UN-Habitat, Islamabad, 04/10). Imam*s* would usually begin their speech with these thoughts:

> I have led [so many] funerals after the earthquake and these people did not die because the earthquake killed them. It was because of their houses that fell on them and killed them. Allah entrusts us with our lives, and we need to protect it by building our houses safely.

Although, the mosque was enlisted as a critical facility[14] for communities in the ERRA CBDRM Programme booklet but it did not receive any kind of support from the government. Likewise, NDMA's National Disaster Response Plans and Provincial Disaster Response Plans mention the role of the mosque only to the extent of dissemination of disaster early warning, since mosques have loudspeakers (National Disaster Management Authority, 2019; Provincial Disaster Management Authority Punjab, 2018).

[14] The concept of critical facilities is to identify those buildings, facilities and services that are essential for people, such as transport, electricity, fire service, hospital and health clinic in the time of extreme emergency (UNISDR, 2009).

3.14 Conclusion

Analysis of five decades of the development planning history of Pakistan has shown that before the 2005 earthquake the country's disaster management remained narrowly focussed on counter-flooding (excluding flash flooding) measures. This focus on counter-flooding was on structural aspects such as the strengthening of embankments and bunds, the digging of new canals, the construction of new dams and the generation of electricity. However, non-structural measures to reduce flood losses, which included the capacity building of communities, related government departments and organisations for flood preparedness, were not emphasised in institutional policy or the structures of disaster management organisation.

Disaster management was highly centralised and skewed towards the response and relief phases of the disaster management cycle, whereas the preparedness, recovery and rehabilitation of the affected people were largely ignored in the pre-2005 earthquake situation. Although there were seven government agencies for response, and six for relief, it was the Pakistan army that would manage a disaster emergency, with subordinate support from civil departments. The legacy of reliance on the army was further strengthened because of the three-decade military rule. This dependence on the army excluded the development of credible civil disaster management institutions.

Before the introduction of the NDRMF, different government organisations, federal and provincial and often with overlapping roles, lacked preparedness, coordination, coherence and a sense of direction on the eve of a disaster. Before the 2005 earthquake, Pakistan did not have a single federal organisation for disaster management with a multi-hazard approach that included earthquakes.

There was minimal allowance for the roles of the private sector, civil society organisations and local community institutions in disaster management before the 2005 earthquake. The fieldwork findings have shown that mosques were only used passively for flood warnings by government organisations. Grassroots government organisations responsible for community level disaster response, such as Civil Defence Departments, were ignored and gravely under-resourced. However, international and national NGOs, local community institutions and the private sector have been playing a strong role on their own, to compensate for the dysfunctional public sector in disaster situations.

Local government reforms were introduced in 2001 during military rule. These reforms affected the pattern of disaster management policy in the aftermath of the 2005 earthquake. As this earthquake was an unusual event in the history of the country, an interim disaster management policy and structure was set up to deal with it. The NDRMF released in March 2007 was prepared with the technical advice of the UNDP Pakistan.

The 2005 earthquake was a lesson-producing event for the whole nation and the government in terms of renewed awareness of disaster preparedness and the mitigation of other hazards beyond flooding. With the promulgation of the Disaster Management Ordinance in December 2006, the country witnessed an apparent paradigm shift from a flood-based and highly centralised contingency disaster risk approach to a multi-hazard and integrated disaster policy incorporating a broad range of stakeholders in disaster management. The NDRMF espoused the aim of engaging with a broad range of stakeholders with the establishment of the central federal organisation. The federal organisation thus set up, NDMA was charged with the whole spectrum of disaster management functions of preparedness, response, relief, recovery, reconstruction and rehabilitation. However, patterns of horizontal and vertical functional overlap, which forces disaster management to work far below the optimum level, among government disaster-related organisations at federal, provincial and district levels persist. Despite the federal government's paper commitment to the cause of reduction in disaster losses and tall claims of mainstreaming DRR in development planning, the real resource transfer to NDMA remains marginal in comparison with the task assigned to the organisation and the actual commitment of the government. The two main federal disaster management organisations, NDMA and ERRA, have come in direct institutional conflict, dimming the hopes of sustainability of community safety initiatives. Over a decade since the passing of the Disaster Management Act 2010, the process of subsuming of ERRA is yet to be matured.

On the same lines, provincial governments are found to be recentralising the powers that were given to districts under the Local Government Ordinance 2001. This recentralisation has direct repercussions on the district government potential of service delivery, including local emergency and disaster management as a whole and particularly the viability of proposed DDMAs.

The disaster management structure continues to be marred by myopic tendencies which override objectivity and hinder the stable progression towards a reduction in vulnerability in disaster-prone Pakistan. Lack of political will and ad hocery on the part of the government are key factors that cast a shadow on the future of disaster management in Pakistan. Overall, fieldwork findings support the view that the paradigm shift in disaster management, though profound in theory, is still elusive in practice.

Given the state of the disaster management institutional policy and structure of the country, the journey to a life safe from disasters seems very long. Meaningful attention and engagement to build capacity at the local level, coupled with resource transfer, appear unlikely to materialise soon. Meanwhile, the poor are still at the forefront of disasters. They continue to live in isolated, mountainous, vulnerable and rural places. However, the beliefs, values, customs and community institutions of the people such as the mosque, become a source of bringing the poor together to stand against day-to-day hazards and vulnerabilities.

Chapter Four examines the actual and potential role of the mosque in disaster management in the aftermath of the 2005 earthquake through a case study in the Khyber Pakhtunkhwa province. This highlights the role of the mosque in generating and promoting linkages among different stakeholders, but also an opportunity to address some of the deeper endemic issues, such as lack of trust and information gaps, that have emerged between the state and the civil society.

References

Ali, I. (2008). Crises of governance. *International Institute for Asian Studies Newsletter, 49*(Autumn), 1–4. http://www.iias.nl/nl/49/IIAS_NL49_0104.pdf

Callisthenes. (1935). Our record of the Quetta earthquake. *The Times*, p. 10.

Cheema, A. R., Mehmood, A., & Imran, M. (2016). Learning from the past: Analysis of disaster management structures, policies and institutions in Pakistan. *Disaster Prevention and Management, 25*(4), 449–463. https://doi.org/10.1108/DPM-10-2015-0243

Cheema, A., Khwaja, A. I., & Qadir, A. (2005). *Decentralization in Pakistan: Context, content and causes*. Kennedy School of Government, Harvard University, MA.

Cheema, A, Khwaja, A. I., & Qadir, A. (2006). Local government reforms in Pakistan: Context, content and causes. In P. Bardhan (Ed.), *Decentralization and local governance in developing countries* (pp. 381–433). MIT Press.

Earthquake Reconstruction and Rehabilitation Authority. (2005). *Earthquake Reconstruction and Rehabilitation Authority (ERRA) ordinance* (G. of P. Earthquake Reconstruction and Rehabilitation Authority (ERRA), Ed.). http://www.erra.pk/aboutus/erra.asp#Ordinance

Earthquake Reconstruction and Rehabilitation Authority. (2010). Community *based disaster risk management (CBDRM)* (E. R. and R. Authority, Ed.; Issue 2010). Earthquake Reconstruction and Rehabilitation Authority, Government of Pakistan. http://www.erra.pk/sectors/drr/CBDRM.asp

Earthquake Reconstruction and Rehabilitation Authority. (2011). *Annual review 2009–2010*. Earthquake Reconstruction and Rehabilitation Authority (ERRA).

Flood Forecasting Division. (2005, January 18–22). Summary of national information on the current status of disaster reduction, as background for the World Conference on Disaster Reduction (WCDR). Pakistan.

Government of Pakistan. (2011). *National Volunteer Movement (NVM): About us*. http://www.nvm.org.pk/AboutUs/Index.html

Government of the Punjab. (2019). *Provincial disaster management authority: DRM institutions*. http://pdma.gop.pk/drm_institutions

Harrison, T. G. (1992). *Peacetime employment of the military: The Army's role in domestic disaster relief*. In U.S Army War College, Carlisle Barracks. U.S Army War College, Carlisle Barracks.

Imran, M. (2010). Sustainable urban transport in Pakistan: An institutional analysis. *International Planning Studies, 15*(2), 119–141.

Jacob, T. (2001). Path-dependent Danish welfare reforms: The contribution of the new institutionalisms to understanding evolutionary change. *Scandinavian Political Studies, 24*(4), 277–309.

National Disaster Management Authority. (2007). *National disaster management framework of Pakistan*. National Disaster Management Authority (NDMA) Government of Pakistan.

National Disaster Management Authority. (2019). *National disaster response plan* (Issue 1). http://cms.ndma.gov.pk/

National Disaster Management Authority. (2021). *National Disaster Management Authority (NDMA): About us*. http://cms.ndma.gov.pk/page/about-us

Planning Commission of Pakistan. (2010). *Government of Pakistan five year plans; Ist plan 1955–60, 2nd plan 1960–65, 3rd plan 1965–70, no plan 1971–76, 5th plan 1977–83, 6th plan 1983–88, 7th plan 1988–93, 8th plan 1993–98 and medium term development plan 2005–10*. http://www.planningcommission.gov.pk/

Provincial Disaster Management Authority Punjab. (2018). *Disaster risk reduction strategy: Provincial disaster response plan*. http://pdma.gop.pk/system/

files/DisasterRiskReductionStrategy-ProvincialDisasterResponsePlan2018%28Final%29_0.pdf

Shah, I., Eali, N., Alam, A., Dawar, S., & Dogar, A. A. (2020). Institutional arrangement for disaster risk management: Evidence from Pakistan. *International Journal of Disaster Risk Reduction, 51*(August), 101784. https://doi.org/10.1016/j.ijdrr.2020.101784

Shaheedi, A. (2021, March). *GB provincial status*. The News International. https://www.thenews.com.pk/print/805077-gb-provincial-status

Siddiqa, A. (2007). *Military Inc: Inside Pakistan's military economy*. Oxford.

Skrine, C. P. (1936). The Quetta earthquake. *The Geographical Journal, 88*(5), 414–428. http://www.jstor.org/stable/1785962

The Dawn. (2009). Rescue-1122 launched in Peshawar. *The Dawn*. http://www.dawn.com/wps/wcm/connect/dawn-content-library/dawn/the-newspaper/national/rescue1122-launched-in-peshawar-189

The Express Tribune. (2019, November 8). *ERRA to be subsumed into NDMA*. https://tribune.com.pk/story/2095704/1-erra-subsumed-ndma-dec-31

The Times. (1935a). The earthquake at Quetta: General's praise of the garrison. *The Times*, p. 4.

The Times. (1935b). The earthquake at Quetta. *The Times*, p. 15.

Thompson, W. C. (2010). Success in Kashmir: A positive trend in civil–military integration during humanitarian assistance operations. *Disasters, 34*(1), 1–15. https://doi.org/10.1111/j.1467-7717.2009.01111.x

UNISDR. (2005a). *Hyogo framework for action 2005–2015: Building the resilience of nations and communities to disasters*. In World Conference on Disaster Reduction, 18–22 January 2005, Kobe, Hyogo, Japan (Issue A/CONF.206/6). United Nations International Strategy for Disaster Reduction (UNISDR).

UNISDR. (2005b). *World conference on disaster reduction: 18–22 January*. United Nations International Strategy for Disaster Reduction (UNISDR). http://www.unisdr.org/eng/hfa/docs/Hyogo-framework-for-action-english.pdf

UNISDR. (2009). 2009 UNISDR terminology on disaster risk reduction. United Nations International Strategy for Disaster Reduction (UNISDR), United Nations Office for Disaster Risk Reduction. https://www.undrr.org/publication/2009-unisdr-terminology-disaster-risk-reduction. Accessed 24 November 2021.

UNISDR. (2015c). *Sendai framework for disaster risk reduction*. UNISDR. https://www.preventionweb.net/files/43291_sendaiframeworkfordrren.pdf

CHAPTER 4

The Role of the Mosque in the Aftermath of the 2005 Earthquake and Its Future Potential

4.1 Introduction

Chapter 3 elaborated on disaster management structure and policy in Pakistan and provided critical insight into the role of key actors in disaster management. Case studies presented in this chapter relate to rural settings in the Khyber Pakhtunkhwa province in the aftermath of the 2005 earthquake. This chapter discusses the two central questions about functional roles of the mosque in response, relief, recovery, reconstruction and rehabilitation phases of the disaster in the aftermath of the 2005 earthquake in Pakistan, and the mosque's potential role in future in similar situations.

The findings are demonstrated in two ways in the Chapter: (1) the different roles of mosques in Banda-1, -2 and -3 in terms of their cultural, psychosocial, economic, social and political dimensions; (2) the roles of mosques are categorised in response, relief, recovery, reconstruction and rehabilitation in the aftermath of the earthquake.

The findings highlight the influence of the mosque in shaping the disaster-risk perception that affects communities' attitude towards disaster preparedness. The findings show both the opportunities and limitations of the role of the mosque. Interaction between mosques and other key actors in the state, civil society and private sector in different phases of the disaster management cycle is illustrated through examples of these interactions from the fieldwork. The mosques' dealing with women are discussed

A. R. Cheema, *The Role of Mosque in Building Resilient Communities*, Islam and Global Studies,
https://doi.org/10.1007/978-981-16-7600-0_4

in a separate section. The role of the mosque as a community-based religious institution is highlighted, and its potential roles, as described by the research participants, are presented. Some parts of this chapter are incorporated from my earlier work during doctoral thesis titled "exploring the role of the mosque in dealing with disasters: A cases study of the 2005 earthquake in Pakistan" at Massey University, New Zealand.

The next section introduces the research sites of primary data collection.

4.2 Introducing the Research Site

Though the study site was district Mansehra, each of the three villages where mosques were located had distinct characteristics in terms of their location, ethnic composition and connectivity. Essential details about each of the Banda are provided below to contextualise the findings presented later.[1]

4.2.1 Banda-1

Banda-1 was located 32 km from Mansehra and there were 250 families in the village. The village was 1.25 km from the main road with on-foot access through narrow hilly passages and a dilapidated bridge (Fig. 4.1). Horses were used to carry foodstuffs and other necessities of life from the road to the village. The village population was scattered in a radius of around 1.5 km. The village had four tribes, namely Swati, Syed, Gujjar and Alali. The members were physically located in separate pockets of the population within the same area. There was no boundary line among the four tribes, but the communities were identifiable based on their clustered houses.

The population of Banda-1 reached 5,000 with the addition of about 150 families (1,000 people) who had migrated from a nearby earthquake-affected area. These people from the earthquake-affected area decided to migrate in search of better economic opportunities.

[1] Some parts of this section draw on my work during doctoral thesis titled "exploring the role of the mosque in dealing with disasters: A cases study of the 2005 earthquake in Pakistan" at Massey University, New Zealand available at http://hdl.handle.net/10179/4080.

Fig. 4.1 Dilapidated bridge to Banda-1

Banda-1 had three mosques in each locality for Swati, Syed and Gujjar, and the Alali community was linked with the mosque in the Swati area. The Alali too had reserved a place for a mosque and bare stones were seen on the proposed construction site. Out of these three functional mosques in the village, none of them was a Jamia[2] mosque. The village people went to nearby Jamia mosques to offer Friday prayers. Friday gatherings brought all males of the communities in an area to the mosque and were therefore major events in which to share and spread information of general interest.

The average land holding size was 4–5 Kanal[3] and Banda-1 land was arid. Water was available in abundance, with a water supply that reached every home. There was a reservoir at a high place, which stored water

[2] The mosque where Friday prayers are offered is called Jamia mosque.

[3] Kanal = 1/8 acre.

from a spring. Water supply pipes ran to the whole village from this reservoir.

In Banda-1, the majority of the people were dependent on agriculture. Farming was subsistence-based but the produce from the land was hardly enough for two to three months after the harvest. The staple included wheat, rice and maize. During FDGs in Banda-1, the research participants were asked about their impressions of poverty and richness. Land ownership was pointed out to be the single most important factor in determining the social profile of a family. Generally, most of the people had their own houses, no matter poorly built. Owing to a simple lifestyle and no connection to the road, there was a low motivation for owning motor transport. Therefore, in the given socio-geographical situation of the village, it was the land that was valued the most by people. In this way, the village people suggested during FDGs that 80% of the people were poor because of small land holdings and lack of a reliable alternative source of livelihood. About 15% could be ranked among middle income while only five percent were thought rich: they owned large tracts of land and had alternative sources of income as well (Fieldwork Journal, June 2009).

The poor class was engaged in casual labour. Most of the poor men had migrated to big cities to find livelihood. They usually carried out low-paid unskilled jobs such as dish washing in hotels. The middle class was made up of those who had land and a permanent alternative source of income.

Banda-1 was vulnerable to seasonal flooding and earthquakes. In 1992, the worst flood had washed away all vegetation on the village land. In the 2005 earthquake, houses across the whole village were damaged and there were 10 deaths. The signs of landslides, which occurred during the time of the earthquake, were visible at certain locations.

The village was remotely located and weakly linked with the rest of the country. Not only was the connection to the rest of the country limited but also there was only one television for every 50 households and two radios in the whole village. I initially thought the poor people of the village could not afford to have a television. However, after having spent time with the community, the research participants discussed another dimension of not having a television at home. It was considered unbecoming of a modest family to have a television. The community did not consider that the television was showing what was religiously and culturally suitable to be viewed. Therefore, it was culturally inappropriate or

Table 4.1 Weak communication links of Banda-1 with the rest of the country

Description	*Number*
Total family units in Banada-1	250
Households with television	5
Households with radio	2
No. of people in government jobs	3
No. of people in private jobs	2

"bigharat", a shameful act to have a television at home (FGD1, Banda-1, 06/2009). This way of thinking refers to the lack of openness of this community as compared to the urban areas where televisions are regular fixtures in common public places such as tea cafeterias. Table 4.1 shows different indicators that reflect the lack of openness and connectivity of the village people with the rest of the country.

Given the above distance between the means of communication, an isolated community and its logical implications for DRR and disaster management, it was important to explore the factors leading to formulations of these invisible societal constructs, which determined what was modest and appropriate for this community. Discussed later in this chapter that the community-based religious institution of the mosque, through messages from the imam, had a significant role in shaping communities' worldviews. Having restricted access to means of communication, such as radio and television, this situation raises serious concerns since these communities are vulnerable to several hazards including earthquakes and seasonal flash flooding and they might not be able to receive a timely warning.

Another socio-economic and socio-cultural dimension of the poor class in Banda-1 was their involvement in environmentally damaging activities such as cutting forest wood illegally. Trees covered the hills around the village and were a natural barrier to hazards such as flash flooding and landslides. Cutting of these trees had a direct bearing on the safety of the village population. One evening when I was coming back along with the locals to the village close to sunset, I saw three donkeys loaded with wood. There were four young men, in their early twenties, with the donkeys. The men were just holding the donkeys and roaming on the side of the path. Surprised to see these people lurking around, I asked my hosts about them. They did not utter anything about them there, signalled through a wink and spoke later only when we had passed them:

> They have cut this wood from the forest, which belongs to the provincial government and now they are waiting for the forest guard to get away from the outer path of the village. Once their companion tells them that the way is clear, they will cross the road from the village, sell it in the market and earn money, which would be enough for a week. After one week, these guys will be on the hunt again. (FGD1, Banda-1, 06/2009)

I asked my hosts whether these people were from the village and they replied "yes". They said they were uneducated, had nothing to do at home in the village and were a sort of a burden on the village resources. The guides and hosts felt sorry for them, and there was a noticeable tacit consent from my hosts by not opposing them openly. Afterwards, when I had built reasonable rapport with his hosts, I asked them why they could not stop these people from destroying the forest. They said the whole village had committed a few years back not to cut trees from the forest. Now, some people had started flouting the common commitment again and no one was taking it very seriously. One reason was that many people were extremely poor, and they had no means of livelihood. Many were those who depended on casual employment and were unable to find regular work. The co-villagers of these subsistence wood loggers were aware of their abject poverty and therefore their activity was considered permissible.

4.2.2 Banda-2

Banda-2 was located 65 km from Mansehra, a drive of about three hours. There were 200 houses in the village and the total population was 800–900. There was a long jeep track connecting Banda-2 to the main road leading to Mansehra. During my travel to the village, several naked hills showed the massive landslides that had occurred during the 2005 earthquake. These landslides badly hampered the response and relief teams in reaching the affected population in the aftermath of the earthquake, the driver for the local NGO informed the author. The road passed in between the winding hills and a river for about two hours, thus access would have been completely blocked after landslides. Even four years after the earthquake, the road was dangerously narrow at certain points where two vehicles could not cross.

One could reach the village by jeep but had to walk to access the houses of the people since there was no road inside the village. The houses were quite spread out and some were physically isolated. The people used fresh spring water for drinking and irrigation of crops. Unlike Banda-1, this village had a *Jamia* mosque which provided a natural gathering place in community settings to talk about common issues such as DRR. A 52-year old imam led the mosque. I attended Friday prayers during my fieldwork in this mosque. A part of the mosque had collapsed because of the 2005 earthquake and the community prayed on the nearby land until repairs were carried out. Until I visited the village, the mosque was still under construction, and thick clothes were hanging in the place of doors to hinder rain and wind.

Once I was in the village, I saw that the Friday speech was delivered through a loudspeaker and it was audible in all parts of the village. There were about 150 men in the mosque listening to the sermon and offering Friday prayers. Due to subsistence agriculture in the village, many of the men worked in big cities or overseas. After the prayers, men gathered in small groups and began chatting, some staying inside and some hanging outside of the mosque hall. The Friday gathering was the biggest weekly social event for the men of the community. The men who would normally be away in their fields on other days would, especially join Friday prayers since it was religiously obligatory and could not be offered except with the congregation. Through these interactions inside and outside the mosque, the author could see the mosque served as a communal place for men to share and update each other. These groups of men discussed and exchanged their views on different issues ranging from the personal health of each other to village, regional and national issues. They would draw up plans to conduct different economic activities for each other such as helping in farming and collecting fodder for animals.

Since there was no provision for women to pray in this mosque, this social space benefitted men directly. However, the women were indirectly connected to the mosque in that they could listen to public announcements and the sermons of the imam, made through loudspeakers, and through men carrying messages back home from the mosque.

Like Banda-1, no NGO was working for the village before the earthquake. It was only local people helping themselves, or occasionally they sought government grants for community projects. During the 2005 earthquake, 37 people had lost their lives and several others were injured. After the earthquake, the village vulnerability and poverty were highlighted, and development organisations turned towards this community with various projects. Despite different projects for community welfare, no organisation, including government institutions, funded the reconstruction of the village mosque. The community on its own raised 900,000 Pakistani Rupees (USD 10,975)[4] for the repair and extension of the mosque.

4.2.3 Banda-3

Banda-3 was located about 9 km from Mansehra linked by a narrow road. It had about 200 houses and the population were 1,000. There was no major damage to this village during the earthquake. The village was situated at the top of a hill unlike Banda-1 and Banda-2, which were in valleys. Similarly, the village had better infrastructure and housing than that of Banda-1 and Banda-2. One obvious reason for this was the village connectivity with the city that had opened several employment opportunities for the people. There were paved streets inside the village. It had two primary schools: one for boys and one for girls. The girls' school building was severely damaged during the earthquake.

The mosque included in this case study was in the centre of the village and was led by an imam who was 45 years old. The mosque was partially damaged in the earthquake. The people continued praying inside the mosque since it was not considered unsafe. However, just like the case of the mosque of Banda-2, this mosque was also repaired and extended after the earthquake, with the community raising 2,200,000 rupees (USD 26,829). No government or private organisation funded the reconstruction of the mosque. The sources of funding included community members who now worked elsewhere in Pakistan or overseas.

4.2.4 Imams and Their Recruitment

The main operator of the mosque in each Banda was the imam. It is important to understand who the imam was, what the process of his recruitment was and what were his major responsibilities. The imam, who had to be a man, was primarily a prayer leader who led prayers five times a day in the mosque. In villages included in the case study, a prospective imam would usually be approached by a village elder. The process would begin with the visit of a village elder to a nearby seminary. The

[4] One United Stated Dollar (USD) = 82 Pakistan Rupee (Pakistan Rupees.) as considered on March 25, 2010.

village elder would see the head of the seminary and place a request for an imam for his area mosque. The head of the seminary would then refer him to any one of his pupils in the area. Once seen and approved by a village elder, the imam would be invited to the village for approval by the community. The community would usually accept and welcome the imam arranged on their behalf by their elder. There was no retirement age for an imam. The imam would continue to serve as long he wished or the concerned community was satisfied with his services.

The imam was not paid a monthly salary in all the three Bandas. People would give charitable donations to the imam twice a year at the time of Islamic festivities. However, this occasional charity was not enough to cover living costs for the imam. Usually, some well-off families in a community would help the imam in cash or kind. In the community of Banda-1, one well-off family was providing food for the imam every day (Male Villager, 35, Banda-1, 07/2009). The imam was found to be an important and respected figure having public faith.

4.3 The Roles of Mosques in Disaster Management

The mosque as a local civil society and community institution contributed to disaster management in the aftermath of the 2005 earthquake. Figure 4.2 shows the roles of mosques in three disaster phases. The sizes of the three boxes in the figure show that the roles of the mosques included in the case studies played were more active in the immediate aftermath of the earthquake than in the later phases when communities began to settle and needed less support.[5] These roles are explained in three phases as grouped in the three boxes in Fig. 4.2

4.3.1 Mosques' Roles During Response and Relief

All mosques in the case studies became initial contact points for initiating response and relief operations in the aftermath of the 2005 earthquake. Many emergency service providers approached mosques to announce

[5] Some parts of this section draw on my work during doctoral thesis titled "exploring the role of the mosque in dealing with disasters: A cases study of the 2005 earthquake in Pakistan" at Massey University, New Zealand available at http://hdl.handle.net/10179/4080.

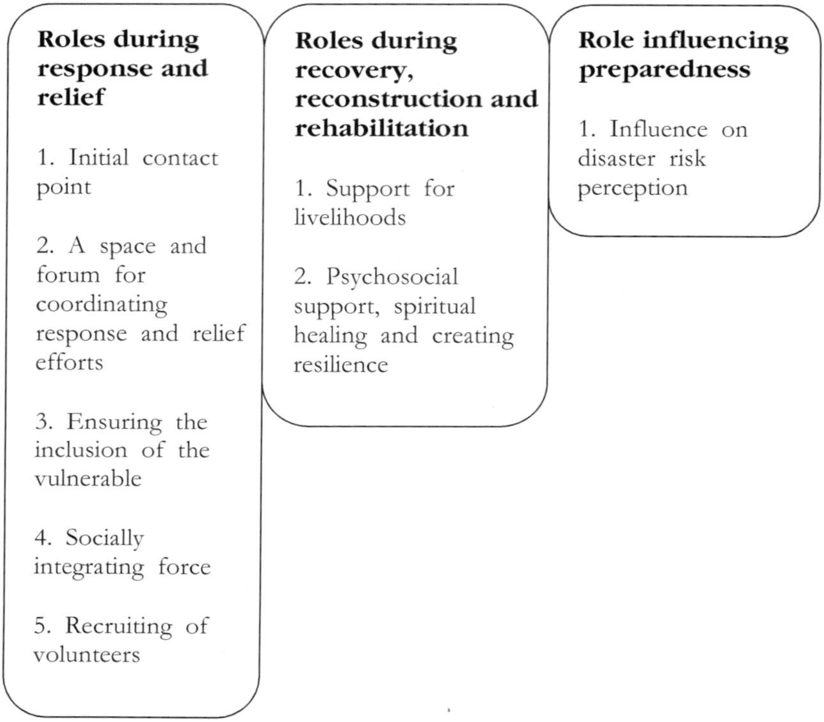

Roles during response and relief	**Roles during recovery, reconstruction and rehabilitation**	**Role influencing preparedness**
1. Initial contact point 2. A space and forum for coordinating response and relief efforts 3. Ensuring the inclusion of the vulnerable 4. Socially integrating force 5. Recruiting of volunteers	1. Support for livelihoods 2. Psychosocial support, spiritual healing and creating resilience	1. Influence on disaster risk perception

Fig. 4.2 Actual roles of mosques in different phases of the disaster management cycle (*Source* Author)

their presence and establish initial contact with communities. One senior officer of the UNDP, who was resident at Mansehra, the case-study district, and who experienced and survived the earthquake, said:

> Most parts of the rural Khyber Pakhtunkhwa, a Pashtun society, used mosques for organising their people after the earthquake. People used loudspeakers for announcing their priorities like removal of debris, motivating the young, telling women to cook for the people busy with debris removal. Within 24 hours, the people of the village knew who had died and who was missing. The whole chain of command was run by using the mosque. (Provincial Coordinator, UNDP, Islamabad, 04/2009)

Eight focus group discussions with men showed that in all the five mosques in the three Bandas (three in Banda-1, one in Banda-2 and one in Banda-3), at a time of frustration and despair in the aftermath of the earthquake, most people turned to the mosques. Mosques proved a spiritually supportive and socially integrating force for the affected communities.

Mosques served as an entry point for response and relief operations in the first few weeks (Country Project Head, FAO, Abbottabad, 05/2009). Later, when the Pakistan Army took over the response and relief drive in collaboration with the UN, meeting points of development organisations and the earthquake-affected community began to shift outside the mosque. However, because of the position of the mosque in the centre of the case-study villages, it served as a coordination place for linking communities with relief. People explained during the focus group discussions that they not only went to the mosque to pray but also to collect food:

> In the start, people gathered in the mosque. For the first three days, there was no food. Then helicopters started dropping food. Then there were recurrent earthquakes, and we used to pray outside. Then the food camp was set up outside the mosque, opposite the mosque. We used to go to the mosque and collect food as well. (FGD4, Banda-1, 07/2009)

In most areas, mosques were physically located in the centre of communities. They were a common and neutral point of contact for communities and because they had loudspeakers, were able to communicate with the whole community present at the grassroots level. In this way, mosques supplied the requisite physical and social space for coordinating and organising relief efforts between the affected communities and relief organisations. Local Muslim relief organisations tapped the potential of mosques in the aftermath of the earthquake as explained by an area head of a Muslim country-level humanitarian organisation:

> From December 2005 to December 2008, 572 mosques were built by our organisation. After the earthquake, the priority was to set up tents for the people. We set up mosques in tents at the start and planned to have a one-room house and we tried to have people not indulge in begging by putting food tables outside the mosque. In regular development schemes, mosques have been major supply routes for accessing locals. The basic

> consultative process was undertaken in the mosque. (District Head Muslim NGO, Mansehra, 04/2009)

He added that they purposefully visited communities at the time of congregation for prayer. After entering the village, their staff would pray with local people first and then engage with them in relief activities. He called this a "natural way" to engage with communities. They would usually have meetings in the mosque and form a disaster management committee:

> In the beginning, when we used to come, we made announcements to gather in the mosque. When our vehicle arrived people would start gathering around us. Also, we planned to arrive at prayer time. People stayed outside the mosque after prayer and we used the natural gathering for prayer for our meetings with communities. To form committees, we used to join congregational prayers and ask locals to join in. (District Head Muslim NGO, Mansehra, 04/2009)

Due to strong family customs of mutual help, which were reinforced by calls from the mosques to look after and help each other to please God, the vulnerable were looked after. Mosques contributed to ensuring that the poor, needy, destitute and vulnerable (children, elderly men and women) were included during the response and relief phase. A male villager, 55, from Banda-1 described his effort to take care of his old mother-in-law:

> I was in Mansehra, buying fruit and vegetables for my shop when felt the earthquake. I reached Bara in an hour. I got down at the bus stop and came to know that three people had died in one house. I asked about my father, met my son and enquired about the in-laws in a nearby village. There was no one in the Mohalla. When all my family members were safe, I went to the village of my in-laws. I took my mother-in-law out from the debris. She survived but her leg was broken. I tied up her leg and put her on a cart and then took her to Abbottabad. There was a great rush in the hospital, and nobody listened to me. Then I took her to a nearby town to a man traditionally trained in binding broken joints.

Such type of community sentiments for the vulnerable was confirmed one evening when I was in Banda-1. There was a meeting at the mosque after the afternoon prayer about leaking pipes and taps in the village (Fieldwork

Journal, Mansehra, July 2009). During this meeting, one villager pointed out that a widow's house at the end of the pipeline was hardly receiving any water because of leakage of water at certain points. A village elder tasked a young man to repair leaking pipes and he was told he should start his work from the widow's home.

Eight of the nine focus group discussions showed that mosques acted as community collection centres for men to share their grief and pain, and then to get organised to carry out different tasks in the aftermath of the earthquake. "The mosque is our village capital", remarked one participant (FGD3, Banda-1, 07/2009). Announcements and sermons from mosques inspired men and women alike to help each other, stay together and be patient. Women also sought guidance and support from the wife of an imam. A male villager (56, Banda-1, 07/2009) explained the situation after the earthquake concerning the social role of the mosque in his village:

> Prayer gatherings grew stronger during the days of the earthquake. The imam recommended that we turn towards Allah. Before and after the prayer, there were discussions and exchange of information including where to find tents. We kept coming and praying in the mosque even though tremors were coming now and then. (male villager, 56, Banda-1, 07/2009)

Mosques in the case-study area and outside also proved fertile grounds for getting resolute volunteers, through religious motivations to please God, to help in response and relief phases of the earthquake. As noted by a team leader of the Food and Agriculture Organisation (FAO) working in the earthquake-affected areas of Khyber Pakhtunkhwa and Kashmir, "mosques can be a good recruiting ground for committed people" (Country Project Head, FAO, Abbottabad, 05/2009). Muslim NGOs called for volunteer community participation with the help of imams by motivating and inspiring people to join in response and relief tasks. This volunteer recruitment was carried out throughout the country where Muslim NGOs collected public donations in mosques and appealed to people to help their organisation by giving them a hand in relief camps in the earthquake-affected areas. Among many such volunteers who replied to such calls for help, I interviewed one based in Rawalpindi, 177 km from Mansehra. He stated that he took leave from his office for 1 month, joined the relief camp of a Muslim organisation at Mansehra and helped the management of the organisation to carry out different tasks (Volunteer worker, Muslim NGO, Rawalpindi, 10/2009). In particular, he worked as a volunteer translator with a foreign team of medical doctors who could only speak English and therefore required a person who could translate the local language into English to help doctors treat patients.

There was also an opportunity to see the role of the mosque in a humanitarian crisis during my fieldwork due to the Swat military operation. This allowed me to relate this situation to that faced in the aftermath of the 2005 earthquake and understand the role of the mosque. Swat is a beautiful valley located 160 km from the federal capital. It is an administrative district of Khyber Pakhtunkhwa. The Pakistan Army launched the Swat operation in May 2009 against armed militant elements supporting the Taliban in Afghanistan. According to the United Nations Office for the Coordination of Humanitarian Affairs, 2.3 million people were internally displaced because of this operation (*Times of India*, 2009). Consequently, the internally displaced person's camps were set up in different districts of Khyber Pakhtunkhwa. I visited three such camps, namely Shah Mansoor camp in district Swabi, Sheikh Yasin camp in district Mardan and the camp set up inside the city church of district Mardan. Public and private sector organisations came forward to deal with this humanitarian crisis.

Donation campaigns were also launched by government and non-government organisations throughout the country to raise funds for the Swat internally displaced persons. I had a chance to witness the attachment of women and men with the mosque, how it worked as a community mobilising institution and the key role of the imam in this regard. One Muslim NGO, Al-Khidmat Foundation, loosely affiliated with Jamaat-i-Islami, which had engaged in the aftermath of the 2005 earthquake as well, came to Banda-1 for fundraising for the internally displaced persons. The *modus operandi* of this team was similar in all the case-study villages. A team of this NGO, comprised only of male members, approached the three mosques in Banda-1. They had their meetings with the imams and community members within the mosques. After explaining their aim and mission, they made announcements from the mosques calling for donations. Men and women responded alike to the calls of action from the mosque. For example, the imams persuaded communities to donate generously to help their Swat brethren. One woman, a teacher by profession, revealed that she donated her jewellery on hearing the call for donations from the mosque for those displaced by the Swat military operation (Female Informant, 36, Banda-1, 07/2009). She stated that she felt her connection with God by giving away her jewellery for this noble cause of helping Swati brethren. All the 13 women interviewed for this study were found to have as profound affiliation with mosques as did the men, although women of Banda-1 were not physically visiting the mosques in their village.

The Al-khidmat Foundation had worked extensively in the earthquake-affected areas. Women were also involved through women-only groups. Working via the mosque helped the organisation to gain legitimacy, acceptance and a favourable reception in communities. Communities then trusted the organisation and provided them with community support. There was a clear distinction between the response to donation calls made through other forums and the response from the mosque. The voice that came through the channel of the mosque carried a unique legitimacy, authenticity and pragmatism. The imam's promise of divine rewards gave it a special appeal. People trusted that donations through the mosque would not be misused. A Muslim NGO calling for donations with the help of the imam's engagement by using the institution of the mosque in rural settings of Mansehra made it a pragmatic and suitable method of collection.

4.3.2 Mosques' Roles During Recovery, Reconstruction and Rehabilitation

At the local level and during the reconstruction phase, there was no broader engagement with the private sector and community institutions to address the local issues of reducing losses from disasters as part of the government reconstruction policy. In the aftermath of the 2005 earthquake, the government monitored public-funded housing to ensure the implementation of seismic-safe building codes. However, ERRA reconstruction guidelines were generic and no micro-seismic zoning had been carried out that could guide private construction in the earthquake-affected areas at the local level (Project Coordinator, DDMA, Mansehra, 05/10). Since no proper seismic zoning was available for Mansehra, private reconstruction was not bound to follow any seismic-safe reconstruction guidelines. In addition, there was no government body to take charge and ensure the implementation of even generic seismic-risk reduction codes on private, small-scale commercial reconstruction such as shopping plazas. Consequently, as observed during fieldwork, community buildings such as mosques were reconstructed without care for seismic reconstruction guidelines in many localities. In one instance, a team of the UN-Habitat intervened to convince a community to build their mosque as per ERRA reconstruction codes, the people disapproved of the suggestion saying "it was not the ERRA's mosque, it was their (the community) mosque" (Social Mobiliser, UN-Habitat, Islamabad, 04/10).

Although mosques' roles differed from each other during the recovery, reconstruction and rehabilitation phases of the earthquake, and some were more active than others, all of those included in my case studies were a source of strengthening livelihoods at the local level. As a community institution and a meeting point for men, they were conducting this role even before the earthquake. However, these mosques provided a continued opportunity for men for chance meetings leading to more organised consultations about crops, casual labour and seasonal work opportunities in other places. This was a critical support to communities in the aftermath of the earthquake once other channels of information (such as the market) were severely disrupted due to the physical destruction of their infrastructure. Once people gathered at the mosque to offer prayers, they socialised, shared information and made livelihood decisions to support each other at the local level:

> Once we gathered in the mosque, we decided whose crop will be watered, cleaned of undergrowth or harvested tomorrow and all the men would go there. (Male Key Informant, 35, Banda-1, 07/2009)

Much of the role of mosques in the earthquake recovery, reconstruction and rehabilitation phase depended on the personality of the imam and his community's perception of his role beyond that of a prayer leader. All the seven imams mentioned in the case studies functioned as facilitators for the efficient use of their mosques. These imams arranged, joined and coordinated meetings between aid organisations and village communities. However, often, the community's perceptions of the role of an imam curtailed his desire for a pro-active role in relief and rehabilitation. Members of a focus group discussion from the Banda-2 community said:

> At the time of the earthquake, some organisations and particularly religious organisations' members would join prayer and call for the help of the imam and other notable community members for distributing relief

> items. However, we do not like our imam engaging in the distribution of relief goods and rehabilitation activities. (FGD1, Banda-2, 06/2009)

When I asked why they did not like it, there was no obvious reason except that historically most of their imams had not engaged in economic activities because their livelihood was provided by the community. Thus, over a period, it was a kind of an unwritten norm for the imam to keep himself limited to matters of worship. The interview with the imam of Banda-2 also reflected the same idea that the community's influence restricted him from engaging directly in relief activities in the aftermath of the earthquake. Rather than distributing relief aid with his own hands to the needy and the marginalised, he preferred to give his input to concerned organisations in the process of identification of the needy and poor in the village. He chose not to lead from the front in this hour of need, respecting the public view of his job (imam, 52, Banda-2, 06/2009):

> I actively engaged with an international Christian organisation's (World Vision) relief and rehabilitation activities. I had meetings with their staff in the mosque where they asked for my help. I helped the organisation's people to reach the poor and the needy in the village. Why I hesitated to join in relief activities directly was because people of the village do not like the imam to engage in such activities which were worldly in their view. Although, from an Islamic perspective, I know there is no such restriction on such activities.

This finding was consistently verified during visits to other communities and interviews with all the seven imams included in the case study. Even the imam who had spent more than three decades in his community did not opt to be at the forefront in the relief and rehabilitation efforts (imam, 55, Rural Mansehra, 07/2009). He stated that owing to his reputation, some of his friends from around the country and overseas directly approached him to deliver relief goods and money among the affected people of his area but he refused. When asked why he did so, he replied:

> There were some people from Gujranwala (a district in the Punjab province) who sent relief goods but those were snatched before organised distribution. So later I separated myself from this process. I declined any such offer. I thought some people might object and think that I have taken a lot of relief items for myself. (Imam, 55, Rural Mansehra, 07/2009)

There was a powerful sense of accountability among imams. They needed to keep their image as public figures above reproach if they wanted to stay in their job with dignity. Imams knew that they had to live in the communities they served. However, I found a contrasting example at a mosque in Banda-3 during the fieldwork and included it in the case study as a comparison with Banda-1 and -2.

The imam of the Banda-3 community not only vigorously and visibly contributed to response and relief phases but still continuously engaged in local rehabilitation projects 4 years after the earthquake. Unlike the other case study villages of Banda-1 and Banda-2, the mosque had a provision for women, who would come for Friday prayers. As the men, women and children of Banda-3 were all linked and affiliated with the mosque, when the earthquake occurred, all the community gathered in the mosque. The imam said that:

> At the time of the earthquake, I was alone in the mosque and busy praying. When I finished my prayer and looked back, there were a whole lot of people including men, women and children who rushed to the mosque to take shelter in the mosque during the earthquake. (Imam, 45, Banda-3, 06/2009)

During the focus group discussion and individual interviews, all research participants reiterated the aim and vision of this dynamic imam as including both religious and non-religious activities. Relief goods were collected at the mosque, and volunteers were mobilised for search and response tasks within the village and nearby places after the imam had delivered motivational calls based on promising huge divine rewards for people's charity.

The Banda-3 imam also headed the Parent-Teacher Council looking after the reconstruction project of the village girls' primary school. Parent-Teacher Councils were formed to restore educational facilities after the earthquake as part of the rehabilitation projects of the USA Aid Agency (USAID). The USAID project-monitoring officer confirmed that they were particularly focussing on imams of mosques to be part of Parent-Teacher Councils during this education project as a matter of policy, although he could not source any document that had such policy in black and white, despite my request (Project Monitoring Officer, USAID, Islamabad, 06/2009). No woman was a member of the Parent-Teacher Council in Banda-3.

During the earthquake in Banda-3 and following days of tremors, the old building of the girls' primary school collapsed so the girls of the village had no place to continue their education. The boy's school building survived the earthquake. The mosque committee headed by the imam made a remarkable decision. They decided to shift the girls to the boys' school and to shift the boys' school to the mosque until the girls' school building was reconstructed. When I visited Banda-3, four years after the earthquake, the boys were still attending classes in the mosque. Instead of sitting at desks, which were stacked aside in one corner of the mosque, they were sitting on the floor and continuing with their classes, while the teachers came to the mosque to perform their official duties. The boys' school in the mosque observed usual working hours, as it would have done in its building. The mosque committee and the Parent-Teacher Council had arranged for the school and prayer timings after consultation with each other so that there was no conflict between the working of the mosque as a dedicated religious institution and the school disseminating knowledge of everyday life.

Later, during individual interviews, community members and the imam of Banda-3 were unanimous in their view about the mosque being a community centre as well as a place for worship. Once, when I was in the mosque during fieldwork, there was a meeting between the Parent-Teacher Council and ERRA to review progress on the reconstruction of the girls' school. I was allowed to join the meeting, which was convened inside the mosque and attended by the imam, members of the Parent-Teacher Council and a representative of an implementing partner NGO. The meeting lasted for an hour and it was surprising to observe the institution of the mosque providing the social space for coordination and collaboration among disaster management actors at the local level. There were 12 people in the meeting and all participants spoke freely about the slow pace of the contractor and other related issues. Everybody, including a regional officer of ERRA and the regional head of the NGO, was sitting on the prayer mats of the mosque. The meeting was presided over by an ERRA officer. The mosque was no less than a coordinating office for the local reconstruction and rehabilitation projects in the aftermath of the earthquake since it brought government, private, and civil society sectors under one roof. Not only this, it was an encouraging environment for the poor, illiterate and marginalised men of the community who raised issues and asked questions from the government officials and NGO representatives during the meeting. Ordinary villagers also questioned the government officials about his responsibility to ensure the timely completion of the school. The government officer made the contractor explain the causes of the delays to the Parent-Teacher Council and other ordinary villagers present in the meeting. He also asked the contractor to commit to a completion deadline. The mosque helped the flow of information in favour of the poor, usually left out, which is essential for improving transparency and accountability of actions of organisations such as government and local and international NGOs. This Banda-3 mosque provided practical engagement with key disaster management actors during the recovery, reconstruction and rehabilitation phase in the aftermath of the earthquake.

Why were this mosque and community different from the other mosques and communities in terms of their engagement in the recovery, reconstruction and rehabilitation phase of disaster management? One of the two main reasons was the personality of the imam. The imam of Banda-3 explained the background to his engagement in the Parent-Teacher Council:

> Before the earthquake, I had continuous interaction with local schoolteachers. Sometimes, I would visit the school during the daytime and have a gossip with them. Also, if any teacher were away on leave, I would voluntarily take over a class. (Imam, 45, Banda-3, 06/2009)

In addition to his engagement with the local community and education sector, the imam of Banda-3 would also not hesitate to see the local councillor, the Nazim (locally elected leader, equivalent of a Mayor) and discuss with him the developments in the mosque. It is the personality of an imam that, to some extent, determines the social role of the mosque and its capacity to perform as a community institution in any non-worship activity such as disaster management.

The second main reason was a community's view of the social role of the mosque that would figure out the limits of the mosque's engagement in disaster management. As seen during the fieldwork, it was the belief of communities that decided the extent of engagement of imams and mosques in all phases of the disaster management cycle. If a particular community wished its mosque and imam to engage in disaster management, it did so. Although an imam had influence, the community and the imam collectively determined the overall nature of engagement and the extent of involvement of the mosque in different phases of the disaster management cycle. In a conflict, it was the opinion of the community that would prevail since the imam stays in his job with the consent of the community. In this way, the mosque's operations were democratic, reflective of the opinion of its custodian community.

The earthquake was a harsh reality that shattered many people's lives, and imams, as popular community leaders, had to have their say in it. Imams, using the platform of the mosque, delivered a strong message that the earthquake was the result of one of two reasons: a test from Allah or punishment for sins. This gave meaning and objectivity to the scenes of unparalleled death and destruction.

Imams preached that communities should weep, repent, ask forgiveness and reconnect and repair their relationship with Allah. The imams were faith healers who led the way to spiritual healing for the affected communities. As a result, the earthquake-affected communities, men in mosques and women at home tried to reconcile their relationship with God through meditation, prayers and supplications, seeking strength to face the calamity. People did not despair, despite feeling cursed by God, and chose to retreat to God as suggested by imams. This is how some of the research participants explained their spiritual reconnection with God in the light of the imams' interpretations of the earthquake:

> Earthquakes occur because of the order of Allah. Our sins are too much. The Prophet Muhammad (PBUH) had informed us fourteen hundred years ago that there would be more earthquakes close to the day of the judgement. As our sins would grow, greater would be calamities. Allah is never unjust. Earthquakes are also a test from Allah. Human beings are a creature of Allah. Allah assesses His people who stay steadfast and patient. (FGD2, Banda-1, 07/2009)

> After the earthquake, the message from the mosque was that Allah was angry with us because of our deeds. Therefore, this punishment has been given to us. Also, it was a test and we must get through this test successfully. (Male Villager, 22, Bara, 07/2009)

Imams were also quoted as saying that all the wealth such as houses, children and their lives belonged to Allah and He could take these back whenever He willed. During focus group discussions, men and women relayed this message from the mosque that afflictions and calamities in this life would be a source of relieving pain and suffering on the day of judgement. This interpretation of the earthquake as a test from Allah and a source of warding off sins for those who would bear it patiently greatly helped the earthquake victims to relieve themselves from psychological and mental sufferings. This psychosocial healing was provided through the creation of shared meaning of life, helped by psycho-religious interpretations by imams through the platform of mosques. The imams referred to the word of God, the Holy Quran and the sayings of Prophet Muhammad (PBUH), as explained in the next section, to back up their speeches. As a result, there was an increase in the number of people attending mosques.

The physical form of the mosque was also perceived as a fountain of divine blessings that provided spiritual healing at a time of great stress in the aftermath of the earthquake. Men received this healing through increased attendance of congregational prayers in mosques, and women through their increased affiliation to and reverence of mosques. This is how Banda-1 women, who never had the chance to physically attend the mosque, felt about the presence of the mosque in their village:

> The mosque is the house of Allah and we must frequently visit it. It has a great obligation on us. By making the mosque live, we will live. If we would leave the mosque unattended, we will be uprooted and ruined. The mosque has innumerable benefits. Because of the blessings of the mosque, we had the great mercy of Allah and were saved from the severe destruction of the earthquake. (Female Villager, 72, 07/2009/Banda-1)

> We are all alive because of the mosque. When there is Azan (call for prayer), then all creations pay thanks to Allah. Only because the mosque is frequented, we would be all alive. (Female villager, 69, 07/2009/Banda-1)

Imams reached out to people to say how they could get back on their feet. The counselling was provided both in public and private spaces. General counselling was provided through sermons, lectures and open talks in mosques and other communal places like markets where Imams met communities. Private counselling was offered when imams led funeral prayers and visited the houses of families who lost their near and dear ones during the earthquake. During focus group discussions, community members acknowledged that it greatly helped them to share their grief over the loss of their loved ones and belongings. Consoling each other by ascribing the earthquake to the will of Allah, as interpreted by imams, helped. The other benefit of this morally contextual spiritual healing was that no instance of theft or stealing occurred in the villages included in the case study even though several houses remained unattended for many days in the aftermath of the earthquake.

The next section discusses some basic implications of the other side of this psychosocial and spiritual healing contribution of mosques in terms of the disaster preparedness attitude of communities through disaster risk perception.

4.3.3 *Mosques' Roles in Influencing Disaster Preparedness Through Affecting Disaster-Risk Perception*

Almost 94% of the research participants (89 out of 94), both literate and illiterate, experts and ordinary persons, urban and rural men and women in this study, responded that earthquakes occur because of the order of Allah. The same percentage of the people told that whichever mosques they attended ascribed the earthquake to one of two causes: a profusion of sins and a test from Allah. It was found in the field-work that imams as opinion-makers played a leading role in shaping the communities' perception of disaster risk. Most of the people living in rural settings of Khyber Pakhtunkhwa included in the case study were found to be deeply influenced by the perception of their imams. Here are a few excerpts from interviews showing how people explained the occurrence of the earthquake:

> Regarding earthquakes, these are [because of] our sins; we should ask forgiveness from Allah. This is His sole discretion to do whatever He wants. What can we do by using a strategy? You can imagine that our area is in the red zone, how long can these buildings stand in the face of an earthquake? (Village Leader, 56, Banda-1, 07/2009)

> There were talks about the earthquake in mosques. Mosques told us all about it in religious colouring like immorality: the interest-based banking[6] system and likewise are responsible for this destruction. (Provincial Coordinator, UNDP, Islamabad, 04/2009)

Imams also influenced the perception of local private sector business people on disaster risk. For instance, a manager of a concrete block factory

[6] The Quran prohibits Muslims from charging or receiving interest on money.

said that his factory dealt with government and private organisations, which ordered blocks for construction (Block Factory Manager, Rural Mansehra, 05/10). There was a rapid growth of the business, as concrete block making became popular after the earthquake given the sudden huge demand for reconstruction of damaged and destroyed infrastructure. One such block factory manager interviewed for this study related to the local mosque of his area. In response to the question on why the earthquake occurred, he told that people's sins were the main cause of earthquakes, as told by the imam of the local mosque in his speeches and sermons.

Due to these impressions, the earthquake-affected people tended to dodge some of the disaster-preparedness guidelines. During a focus group discussion with the Alali community in Banda-1, the research participants told that they had no respect for the building by-laws during the construction of their new houses. Since these people had migrated from another place, they were somehow able to escape the monitoring system of ERRA. It was suggested by the key informant that they received official aid from their original place but constructed their houses here. Why did these people not follow seismic-safety building codes, even though they had suffered the worst earthquake of their life, and lost lives and property? The imam of the Alali community, whose visit to Banda-1 coincided with my fieldwork time, living in the city, took part in a focus group discussion and said that:

> One can follow the different suggestions by various organisations about the construction of our houses. However, you can see for yourself that these lofty mountains were moved because of the earthquake so what do you think of these small houses. Did you not see the plaza in the middle of the rubble of Balakot, which did not fall during the earthquake? It was only because the owner of the plaza was pious and paying the Zakat.[7] (Imam, 55, Rural Mansehra, 07/2009)

Communities included in this study challenged the rationale of disaster preparedness. In particular, the illiterate contrasted the idea of disaster preparedness with the absolute link of earthquakes with divine punishment for an excess of sins. When I asked religious scholars about it, they confirmed the religious interpretation that correlates the occurrences of

[7] This refers to a compulsory annual deduction of 2.5% from the wealth of rich Muslims for redistribution among the poor and the needy.

natural calamities including earthquakes with the deeds of Muslims and trials from Allah. A prominent local religious scholar who was the imam of the central mosque of his village and head of an educational institution remarked:

> Being Muslims, we will look into The Quran and Hadith.[8] Earthquakes occur because we commit sins do not fulfil the rights of the people. These, including earthquakes, droughts, or more are punishments from the heavens. The other factors scientists tell us that may also be the reason, I do not reject them, but the real cause is our sins. If we end the cause, then earthquakes will not come. Islam has allowed us to follow precautionary measures so we can do that. I believe in the wisdom of carrying an umbrella if rain is expected but we need to keep in view the capacity of an umbrella, which may be exceeded by the power of the wind or a storm. Look at the example of Hurricane Katrina in the USA that had far better preparedness measures in place than ours but the superpower could not cope with the devastation and suffered huge human and financial losses. (Imam, 45, Rural Abbottabad, 04/2009)

It is important here to look at the source of knowledge that feeds the mosque and imam about their orientation towards natural hazards, such as earthquakes. When consulting the Holy Quran and Hadith, the two authentic sources of Islamic law for Muslims, there are several references to the occurrences of natural calamities including earthquakes. There is a full chapter in the Holy Quran (Chapter 99; 1–8) titled as the earthquake:

> When the earth is shaken with its (final) earthquake. And when the earth throws out its burdens. And man will say, what is the matter with it? That day it will declare its information (about all that happened over it of good or evil). Because your Lord will inspire it. That day humankind will proceed in scattered groups that they may be shown their deeds. So, whoever does good equal to the weight of an atom (or a small ant) shall see it. And whoever does evil equal to the weight of an atom (or a small ant) shall see it.

In addition to this mention of the earthquake in this chapter on the day of judgement, there are several other references in the Holy Quran, for

[8] It refers to the teachings of Prophet Muhammad (PBUH) taught by him to his companions and considered as a source of legislation along with the Holy Quran in Islam.

example, narrations about Prophets Moses (PBUH) and Noah (PBUH), which relate to the blessings of Allah for obedient people and punishment for renegades. In these stories, the renegades were punished through unusual types of heavenly disasters, natural. This theme of punishments through natural hazards, including the 2005 earthquake, appeared in several discussions with research participants, and particularly with imams. A clear link between the bad deeds of people and natural hazards is provided in this narration attributed to the Prophet Muhammad (PBUH) (Al-Tirmidhi Hadith 5450, Narrated by Abu Hurairah):

> When booty is taken in turns, property given in trust is treated as spoils, Zakat is looked on as a fine, learning is acquired for other than a religious purpose, a man obeys his wife and is un-filial towards his mother, he brings his friend near and drives his father far off, voices are raised in mosques, the most wicked member of a tribe becomes its ruler, the most worthless member of a people becomes its leader, a man is honoured through fear of the evil he may do, singing-girls and stringed instruments make their appearance, wines are drunk, and the last members of this people curse the first ones, look at that time for a violent wind, an earthquake, being swallowed up by the earth, metamorphosis, pelting rain, and signs following one another like bits of a necklace falling one after the other when its string is cut.

Common rural people were often found to relate any of the causes mentioned in this saying of the Prophet Muhammad (PBUH) as the prime reason for the earthquake. A disabled male villager who was also illiterate remarked that non-payment of Zakat by the rich was the main reason for the occurrence of the earthquake (Male Disabled Villager, 60, Banda-2, 06/2009). The question arises that in a country where half the population illiterate, how are these messages from the Holy Quran and Hadith that affect the perception of disaster risk of the general masses, transferred to most people.

In the case study, religious leaders, including imams, were the principal source of transfer, and the mosque was the most important primary medium of transfer of religious knowledge to communities. Research participants in focus group discussions referred to a public talk by a famous scholar (FGD1, Banda-1, 06/2009). I was persuaded to buy a compact disc and listened to the speech. This was a speech by Tariq Jamil, a national-level scholar, popular across the country owing to his persuasive eloquence. This scholar visited the earthquake-affected areas and delivered public talks to gatherings of thousands, referring to this traditional narration of the Prophet Muhammad (PBUH) to explain the background causes of the earthquake. Later, his speech was aired on FM radio. One point in his speech, which concerned how he connected earthquakes with religious interpretation and distanced them from scientific evidence, reflects his stance. He stressed that science might explain the reasons for the earthquake, for instance, that tectonic plates moved but could not answer questions such as why the plates moved. The scholar emphasised that it was the sole discretion of Allah to order the moving of tectonic plates. The visualisation of these impressions was seen in the fieldwork. A village elder of Banda-1 was fully aware of the location of his village in the red zone, which was highly vulnerable to earthquakes, yet he challenged the wisdom of using any risk-mitigation strategy: "How long can our buildings stand during an earthquake if Allah wants them to fall"? (Village Leader, 56, Banda-1, 07/2009).

Only one imam (out of seven) interpreted the Quranic warnings of natural hazards, including earthquakes, differently. A university lecturer and an ex officio imam said that the 2005 earthquake was related to the movement of tectonic plates and thus was not a punishment of deeds. He proposed that the hereafter was meant for settlement of good and bad actions (University Lecturer, Muzaffarabad, 05/2009). However, the other six imams found clear religious linkages between actions of people and natural hazards. A higher-ranking government officer and an ex officio imam of a government-controlled prominent mosque in Islamabad had a critical view of the role of the mosque. He said natural hazards could be interpreted in two ways: religious and scientific. He suggested looking at them from a cause and effect point of view. He remarked that floods and earthquakes are "lesson-producing events" and we should recognise our responsibilities and try to do our best to minimise their effects beyond mere repentance from sins. Underscoring the role of every relevant institution during a calamity, he stressed that the mosque should play its role, which it was not then playing:

> There are more than 300,000 mosques in Pakistan and a huge amount of money is donated to them but less than five per cent have a welfare orientation. (Director General, Dawah Academy, Islamabad, 06/2009)

Most of the people from the affected communities had an overriding belief that the earthquake was due to the will of Allah. Overall, risk perception was not a function of chances of an impending disaster alone but was also shaped by a set of socio-religious factors. Piety was considered a parallel survival strategy and even in some cases precluded adherence to seismic safety building codes. Thus, the people in case-study areas did not show willing compliance with earthquake-resistant reconstruction codes introduced by ERRA. Only those owners who received assistance from ERRA built their houses as per the building by-laws since it was a pre-condition for assistance and there was a vigilant monitoring mechanism as well. Those who had a chance not to abide by the by-laws constructed their houses in legally prohibited and earthquake-vulnerable styles. Mosques through imams were the primary source of setting parameters of risk perception in communities.

There was a parallel interpretation of disasters in terms of a punishment or a test from God and geological reasons, thus providing room for disaster preparedness. The former half of this interpretation of coping with disasters through the strength of virtues, on the other hand, was far more deeply embedded in communities' minds than the physical explanation. It was this physical explanation that could prepare people to take practical steps for disaster preparedness. The overwhelming inclination of communities towards their morally contextualised spiritual reformation as a parallel disaster coping strategy had direct implications since the level of perceived risk was fundamental to preparedness. These fatalistic tendencies lead to a compromise on safety standard advised by the government, which might result in an increase in preventable loss of lives in case of a future earthquake.

Imams were the mouthpiece and main channel of transformation of religious interpretation of natural hazards to ordinary people. They drew references from the Holy Quran and Hadith to back up and propagate the religious interpretation of the 2005 earthquake. This interpretation had a strong moral context. In addition, it negatively influenced people's orientation towards disaster risk perception and precluded some practical measures towards disaster preparedness. Although this interpretation promoted fatalistic tendencies in people, it did not turn them completely fatalistic as they participated in disaster-preparedness initiatives from time to time.

The next section looks at the interaction of mosques and women before and in the aftermath of the 2005 earthquake.

4.4 Mosques and Women in the Post-2005 Earthquake Period

Before the 2005 earthquake, the mosque was instrumental in promoting selected development activities, such as stitching, sewing and handicrafts, among women[9] (FDG2, Banda-1, 07/2009). Although these activities would be coordinated through the mosque (loudspeaker announcements and approval/support of imam), they would be located in someone's house, not in the mosque. Also, 46% of the female research participants (6 out of 13) were literate because they received religious education inside mosques and their attached seminaries. Religious education was accessible, free of cost and supported by parents, whereas it was difficult and costly to send girls to far-off schools. Until the restoration of the mosque infrastructure damaged due to the earthquake, 150 girls were studying (up to primary school level) in a nearby (Banda-1) Islamic seminary with boarding facilities (Imam, 75, Rural Mansehra, 07/2009).

After the 2005 earthquake, women were connected to the mosque through loudspeaker announcements. In some cases, they were also involved with the mosque through the wife of the imam who would teach them the Holy Quran and disseminate information. Women were not physically seen in mosques and this was confirmed when I attended prayers in Banda-1, Banda-2 and Banda-3 mosques and spent time with communities but could find no women. An old woman explained her attachment to the mosque and the importance of the mosque, even though she could not attend prayers there (Female Villager, 72, Banda-1, 07/2009):

> We do not go to the mosque because it does not have a facility for women. We hear Azaan and prepare for prayer, our children go to the mosque and learn the Holy Quran and our men go there and offer prayers. The mosque is a great blessing for our village.

Keeping in view this separation of women from mosques in many places, a research participant cautioned "the mosque as a forum is okay to be used in disaster management but who will care for women, there is a fear of

[9] Some parts of this section draw on my work during doctoral thesis titled "exploring the role of the mosque in dealing with disasters: A cases study of the 2005 earthquake in Pakistan" at Massey University, New Zealand available at http://hdl.handle.net/10179/4080.

cutting off half of the population" (Country Project Head, FAO, Abbottabad, 05/2009). The study found two main reasons for this separation of women from the mosque: the community did not like it, and mosques did not have an elaborate facility for women. Although it was not a priority for the village men to build a provision for women, they also could not free up enough resources from other more pressing needs to build a facility for women to pray in the mosque. In addition, the local customs concerning purdah (veil) did not encourage women's broad participation in public life. There was strong gender separation in Banda-1. It was not usual for women to visit the market except for an explicit purpose such as to see a doctor. So, often men would be seen in markets in Banda-1 and nearby towns. Most of the women, when seen in the market, were observing purdah. This veil would hide their body and face in loose clothing, called a chador. I observed many times on my way to the village that women would give way to men by moving towards the edge of the path. They would not exchange greetings with men. In the same way, they would not go outside unless accompanied by a close male family member or an old woman.

Furthermore, women had two other religious limitations restricting their regular attendance in the mosque. First, as it was optional for women from the religious perspective to pray with the congregation in the mosque, so they could not insist on visiting the mosque five times. Second, the mosque was a religious restriction, physical in nature, and they were not to visit the mosque during the time of their menstrual cycle and even not pray at home.

It was observed that religious interpretations were held by the majority of men and women alike and deemed culturally appropriate and valid in all forums of community decision-making. For example, there was a strong spiritual and symbolic connection between women and the mosque: "had there been no mosque in the village, we would have been ruined", was a recurring thought in women's interviews (Female FGD, Banda-1, 07/2009). As with the imam who attributed the survival of a left-over plaza in Balakot after the devastating earthquake to the piety of its owner (Imam, 55, Rural Mansehra, 07/2009), a woman attributed the destruction of a nearby village, which was razed to the ground, to the absence of any mosque there (Female Villager, 44, Banda-1, 07/2009).

In addition to this symbolic connection, female participants effectively perceived the physical structure of the mosque as an institution and tool of survival against the earthquake. Revealed more in women's dialogues than in men was the use of the mosque as a shelter during the earthquake and the following tremors. The majority of the female research participants, 11 out of 13, considered the distinctive importance of the mosque as a critical place for shelter during calamities, including earthquakes. This choice was quite different from ERRA's vision, which considered schools as critical emergency shelter centres. However, this choice might be partly due to the general perception that government buildings including schools were poorly built, given the widespread corruption in the public sector. The women stated that they rushed to the mosque for shelter during the aftershocks. The concept of the soundness of the mosque building did not play any role in their decision to seek shelter there. The women thought that the mosque was the house of Allah, built by communities, and therefore had divine protection against forces of nature such as the earthquake. Not only would this protect the earthquake aftershocks, but also death in the mosque, in the worst scenario, would help in negotiating their pardon in the court of Allah on the Day of Judgement. This clearly showed the degree of sacredness attached to the mosque, considered safer than the government-built community school, by the female research participants of Banda-1: they said they would like to embrace death, if they had to, in the mosque. This type of death was deemed equivalent to martyrdom, a highly cherished ending of life in the Muslim faith.

The women of Banda-1 were asked about the most relevant and important things regarding their village to them in addition to the mosque, they came up with almost the same list of items as men: bridge, roads and hospitals. However, this consensus of men and women did not mean the same level of participation in public life. Owing to purdah, the women would meet in women-only groups and not join the men in the disaster management decision-making process. All meetings and gatherings in mosques involved men alone. As per the existing norms of the area of Banda-1, Banda-2 and Banda-3, women were neither invited nor supposed to join meetings between development organisations and community men in the mosque. However, they would stay updated on all the developments in their village through their men, their children, the Imam's wife and the loudspeaker announcements.

In the aftermath of the earthquake, during the recovery, reconstruction and rehabilitation phase, a protocol was in place through which local development organisations accessed women even although they did not physically come to mosques. The Field Coordinator (Field Coordinator,

SUNGI, Mansehra, 06/09) stated that his organisation would send a women-only team to Banda-1. Respecting the local culture, the team would be dressed in a culturally appropriate way and it would approach only village women. Usually, the women would gather in a designated house, mostly the house of the village chief, after hearing an announcement from the mosque. The mosque, through the imam and other community members, was to function as a gatekeeper, so the women could feel comfortable about the proposed activity of the NGO in the village. During this meeting in the mosque, any cultural gap would also be bridged by discussing the *modus operandi* of the NGO. This kind of protocol to engage women was also confirmed by the district head of the Muslim NGO, Al-Khidmat Foundation. He explained that women had their separate establishment and system of participation in development activities (District Head Muslim NGO, Mansehra, 04/09): the women's section of their organisation that would contact local women to form women committees. Some of these committees in urban areas were set up to address women's needs during the earthquake recovery and rehabilitation. Men would not form part of women's committees but would provide logistics such as the setting up of offices and transportation of goods.

In the Community-Based Disaster, Risk Management programme explained in Chapter 3 the imam's engagement was critical for the involvement of women in the training programme. I noted during the programme that in a classroom, the women were sitting behind men with a gap of two rows of chairs between. The master trainers informed me that they arranged separate training sessions for women in some cases where village people did not like their women seated side by side with men. Since this team had worked in Mansehra before moving to Abbottabad, they mentioned one such case in Banda-2. The women of Banda-2 requested additional women-only training sessions beyond those normally provided. The female master trainer stated that more women developed interest in learning new life-saving techniques once the first batch was about to graduate in Banda-2 (Fieldwork Journal, May 2010). In a group discussion with female participants at the end of the training, all the participants were enthusiastic and enthusiastic having received this six-day training. A young female participant, on being asked about the utility of the training for women, explicitly stated that:

> First aid training was more important for women than men since it's women who stayed at home all day along with small children while men were outside to earn livelihoods and, therefore, they were to handle any contingency on their own related to household injury, fire, earthquake or anything else. (Fieldwork Journal, Rural Abbottabad, May 2010)

Outside the humanitarian situation, some local development organisations collaborated well with women. These organisations used female staff and connected with women. A male villager revealed in a focus group discussion that the female staff of a Christian hospital operating in a nearby town had imparted midwifery training to the village women (FGD2, Banda-1, 07/2009). He stated that his wife received this training and had assisted in 20 births since then; obviously, this was a great help to an isolated village community who had to carry every seriously sick person needing a doctor on their shoulders 1.5 km through a narrow, hilly passage before getting to the main road. Likewise, a local NGO had given 25 days of stitching training to the women, and another organisation had loaned money to female households.

Overall, these examples of women's involvement in training and development work show that women, like men, have understanding, potential and enthusiasm to contribute to DRR and community development when provided the opportunity in a culturally and religiously appropriate way. They also had a strong affiliation with the mosque. The mosque was a sheltered place for them and their children in case of any future earthquake. On spiritual grounds, the mosque had a divinely iconic value for women and was seen as a blessed place critical for the existence of the village during disasters.

This section has shown that women, although rarely seen in public, participated in the earthquake recovery, reconstruction and rehabilitation phase. However, these activities took place outside the physical space of the mosque because of local purdah customs but also because of religious limitations on the physical presence of women in the mosque. Therefore, women were not invited to any disaster management meeting inside a mosque; rather, they would be informed through an exchange of information from men, children and loudspeaker announcements from mosques. The findings show that the mosque itself as a physical space for women's engagement in disaster management was not available but its support as a forum was critical for their engagement outside the mosque.

Mosques were used by imams and men to promote and facilitate religiously and culturally appropriate development and disaster management activities among women.

The next section discusses the interaction of other key actors with the mosque.

4.5 Interaction of the Mosque with Other Key Actors

This section examines the interaction of the mosque with the government and the private sector during the recovery, reconstruction and rehabilitation phase of the earthquake.[10]

4.5.1 Government

As discussed in Chapter 3, religious institutions have generally been ignored as a stakeholder in disaster management, and ERRA did not choose to engage with mosques in the capacity of a community institution as a matter of policy. However, the mosque made a notable contribution to realising the goals of ERRA in many ways. It served as a media centre for advertising the messages of ERRA to those affected by earthquakes. When I asked the senior-most officer of the organisation, the chairman, about the role of the mosque in ERRA's operations, he said:

> We used mosques for public announcements since the mosque is a collection point. For example, to distribute different kinds of undertaking forms, we used local institutions according to local priorities. Sometimes it happened to be a dispensary and on another time a shop. (Chairman, ERRA, Islamabad, 05/2009)

The ERRA Chairman also said, "we have used mosques but very informally" (Chairman, ERRA, Islamabad, 05/2009). He explained that committees were formed for the verification of damage claims. Besides government officials, local notables including teachers, village elders and

[10] Some parts of this section draw on my work during doctoral thesis titled "exploring the role of the mosque in dealing with disasters: A cases study of the 2005 earthquake in Pakistan" at Massey University, New Zealand available at http://hdl.handle.net/10179/4080.

imams served as members of these committees. ERRA neither formally engaged nor prioritised mosques as community-based local partners in any of the disaster management cycle phase. This was in contrast to some international donors such as USAID and UN-Habitat (explained in Chapter 5). The USAID explicitly chose to engage with mosques and imams during the rehabilitation of educational facilities (Project Monitoring Officer, USAID, Islamabad, 06/2009).

The mosque was too important a community institution to be bypassed. Though not formally written as a guideline, and perceived as informal by the chairman, ERRA was connecting with the mosque on the ground to carry out disaster management tasks just like other development partner organisations. ERRA field officials adopted and respected local customs and were seen conducting government business in the mosque with the community in Banda-3. Most of the research participants from Banda-1, -2 and -3 confirmed that several ERRA operations, particularly in the beginning, were conducted and routed through the mosque, especially announcements and community meetings.

The official ERRA CBDRM booklet explaining the aim and design of the formulation of Union Council Disaster Management Committees explicitly stated that the religious leader should be part of any typical Union Council Disaster Management Committee. Similarly, Union Council Disaster Management Committee guidelines enlisted the mosque after the school and hospital in the list of critical community facilities. The basic information dossier of Union Council requires the entry of details about the mosque, including the number of worshippers coming to the mosque regularly, the maximum number of people the mosque could accommodate, and the structure of the mosque building—whether mud/straw, concrete or reinforced concrete. A schoolteacher cum ex officio imam of the village, along with other representatives from different local government departments, including education and health, were participating in the training. All these people, including women, received training about firefighting and first aid with other community members in a school building belonging to an isolated and mountainous Union Council of the district of Abbottabad. Besides, the two master trainers were found to be ensuring the inclusion of the imam in training programmes. This was not only to conform to the instructions given in the ERRA booklet but also to be received well in the community. As stated by the master trainers, imams' views usually have a strong bearing on the response of communities to outside development organisations.

It was important to win the trust and confidence of the imam for the success of the programme in terms of community participation in particular and support in general. As mentioned in Chapter 5, the master trainers had learnt the art of integrating religious knowledge with technical knowledge for promoting development, relating the Noah's Ark story in The Holy Quran to disaster preparedness, to perform their job by winning the trust of communities in their social space (Fieldwork Journal, rural Abbottabad, 05/2010). I asked the imam about the participation of women in training and he said, "women should be included in the training as they would accompany Muslims in battles for nursing the wounded at the time of Prophet Muhammad (PBUH)" (Fieldwork Journal, rural Abbottabad, 05/2010). According to master trainers, it was essential to win the support of the local imam, in particular, for the inclusion of women in the programme.

Although the men and women considered the mosque as a repository of their faith, culture and civilisation, the mandated government agency for reconstruction, ERRA, did not build or repair the damaged mosques. The people responded in the same way as they had to a team of UN-Habitat who tried to convince a community to build their mosque as per seismic codes at another place: "it was not ERRA's mosque, it was their (the community's) mosque" (Fieldwork Journal, rural Abbottabad, 05/2010). Mosques were rebuilt by communities relying on traditional masonry after the earthquake without any regard for earthquake seismic-safety construction codes. Since the government stayed away from mosque affairs and did not support their reconstruction, it did not have any control over the way mosques were rebuilt. However, the government facilitated some donors interested in rebuilding mosques, mainly in urban areas, such as the reconstruction of the District Headquarter Complex mosque in Muzaffarabad city by the Turkish Government (Earthquake Reconstruction and Rehabilitation Authority, 2011).

The women's perception of the mosque and the public policy to stay away from mosque matters have a direct bearing on the success of any disaster management policy at the local level. The consequences of this could be deadly for those seeking shelter in the mosque in case of a future earthquake, most likely to be women accompanied by children. Their homes may have far better survival probability than the mosques since their houses were built with the assistance of the government, following ERRRA seismic building codes.

Thus, the question arose as to why the government was consciously and meaningfully hesitant to appreciate their contribution and engage with mosques. Research participants had different apprehensions including disaster perception, an unknown fear of not having done this before, conservatism and sectarianism. Hardly, anyone in the government denied the tangible contribution of the mosque, yet they had abstract presumptions, not always based on evidence, about engagement with the mosque.

The biggest obstacle for the government to the inclusion of mosques in disaster management was the mosque's traditional and non-scientific approach to understanding disasters. The chairman of ERRA thought it was very difficult to ask a person to be involved in disaster preparedness when they had a belief that everything was going to happen on the order of Allah. The ERRA Chairman was referring to the typical character of an imam who would usually ascribe earthquakes to divine afflictions or tests. In this context, it might not have been an efficient choice to engage with mosques. Given the paucity of time and resources with ERRA, "we went for the least resistant and most cost-effective way of intervention", the chairman remarked (Chairman ERRA, Islamabad, 05/2009). According to the chairman, the mosque did not appear to be an institution to provide a quick-fix solution after the earthquake in the view of the government.[11]

Another reason for neglecting mosques in the disaster management process was their perceived conservative interface towards development. The head of a provincial chapter of ERRA perceived imams as traditionalists who would not welcome progress in society (Director-General Provincial Reconstruction and Rehabilitation Authority, Abbottabad, 05/2009). He acknowledged the importance of mosques but emphasised that:

> A mosque is the epicentre of our society. Generally, very poor people go to the mosque and people of this area [Khyber Pakhtunkhwa] are very religious and they listen to Moulvi.[12] Definitely, Maulvis [plural of Maulvi]

[11] Although the government had limited experience in dealing with seminaries, which are usually attached to mosques, there was hesitation to engage with mosques as such, as a matter of policy in the aftermath of the earthquake. The government had engaged with seminaries in General Zia's regime (1977–1988) and again tried to engage under General Musharraf (1999–2007).

[12] It is a local informal title for imam, sometimes considered offensive.

> are our stakeholders but we do not involve them formally. They only tell people that earthquakes occur because people are sinners and immoral. Mosques are an important community mobilising institution, but they are not playing their due role.

Among other apprehensions, people were concerned that engagement with religious institutions like mosques could result in discrimination based on sectarian affiliations. Imams were questioned about their views on sectarian differentiation during the rescue, relief and rehabilitation processes in the aftermath of the 2005 earthquake. All the imams were against any kind of discrimination, including on a sectarian basis, while helping people during disasters. No such instance occurred during this research to suggest otherwise. This could be due to sectarian harmony and also because the majority of the population of areas under study belonged to one sect (Sunnis) and therefore the issue of sectarian friction was not relevant.[13]

Similarly, a senior officer of ERRA, dealing with social sector programmes for the earthquake-affected areas, stated that they were told to stay away from mosques in disaster management (Deputy Director-General ERRA, Islamabad, 04/2009). This officer, who had more than 10 years of experience of public service in district management across the country, commented on the relationship between the state and the mosque: "We (the state) never felt the need to establish links with the mosque as an institution and we never dared to do so". There was a tacit hesitation and fear of the unknown about formally acknowledging and engaging with the mosque, though it was happening informally. Besides other socio-political factors, this situation indicated the lack of research and awareness regarding the role of religious institutions in disaster management in particular and development in general.

Overall, there seemed to be a large gap in the understanding and perception of the role of the mosque and the imam between those stationed in headquarters like ERRA and those on the ground (like the CBDRM master trainers and Un-Habitat social mobilisers). The field staff used mosques to win community support and network with other actors. Specifically, mosques were useful institutions with which the government engaged, although not initially intended because they were there on the ground. The government used mosques for advertising its policies and

[13] Sectarian clashes usually occur between Sunnis and Shia.

plans. However, mosques were used far less than their potential value due to unresearched concerns (e.g. conservatism and sectarianism). This highlights the lack of research about finding an innovative approach through building synergy about the role of the mosque in particular and religious institutions in general in disaster management.

4.5.2 *Private Sector*

At the local level, the mosque was a meeting and communal place for private sector people of Banda-1 local market. Since this mosque was fulfilling the religious needs of the local business people, the local market people maintained and looked after it. The earthquake damaged this mosque, and a local saw machine owner offered his business place for prayers. The mosque was shifted to this business courtyard until the mosque could be repaired. Unlike schools and hospitals, mosques did not stop functioning, even for a single prayer, because of the destruction of their bricks and mortar in the earthquake.

Some of the local, private sector entities used the centrality of the mosque and the cooperation of the imam as a strategy for carrying out their business with community support before and after the earthquake. For example, underground electric water pumps were a major source of drinking water for the rural communities of the twin districts of Abbottabad and Mansehra. One water pump government contractor based in Abbottabad but also working in Manshera reflected on his experience of engagement with the mosque before and after the earthquake. He stated that his company would get a work order to repair an underground water pump from the local government works department (Government Contractor, Abbottabad/Mansehra, 05/10). This work permit might relate to the repair of a pump in a remote rural community where they had never been before. The contractor stated that his team would ask people in the street about the location of the underground water pump after entering the village. The contractor identified that the other main strategy would be to approach the mosque in the village since the minarets of the mosque are usually visible from a distance so it is easy to locate. Besides, most of the time, water pumps were located close to the mosque in the centre of a village. Another reason to approach the mosque and the imam was to request volunteers. If the contractor needed volunteers to help his team, the imam would announce in the village through the mosque and people would come forward to help them. He further added that even the imam would offer his labour for help in some cases. The contractor worked on the same lines in the aftermath of the earthquake, using the centrality of the mosque for finding the water pump and the imam's support for seeking community volunteers.

When food supplies were directly distributed among affected communities at the time of the response after the earthquake, mosques were involved. However, during the relief phase, the mosque facilitation role was phased out once those affected had shifted to makeshift camps and humanitarian organisations had arrived and started providing different services to the affected people in a formal way. The managing director of a Mansehra-based transport company stated that his company would then directly deliver the supplies to warehouses established by humanitarian organisations in earthquake-affected areas (Managing Director, General Supplies and Transportation Firm, Mansehra, 05/10).

Private sector community members provided direct support to some imams after the earthquake. For example, there was a small mosque in the locality of five or seven houses, and the parents of the imam died, and his house was destroyed in the earthquake. A rich local bulk trader and other locals helped the imam financially and supported him (Bulk trader, rural Mansehra, 05/09).

The mosque was found to be a silent but important local partner of the international private sector organisations. For example, the project officer of an international multi-disciplinary consulting firm, carrying out third-party validation for Islamic Development Bank, stated that the firm's field supervisors immensely benefited from mosques during the verification process of housing reconstruction in Khyber Pakhtunkhwa (Project Officer MM Pakistan, subsidiary of Mott MacDonald International UK, Islamabad, 04/10). During this process, the field supervisors frequently sought the assistance of imams to introduce them to the community, and to be allowed to use the mosque as an office and for overnight stays while in far-flung areas like the district of Kohistan. Although the firm also collaborated with local NGOs, the assistance of the imams was unique. In the context where there was strict compliance of purdah customs in some areas in Khyber Pakhtunkhwa, it was exceedingly difficult for outside people to talk to women when men were not found at home during the time of verification. As the men would usually be away earning a livelihood in big cities like Karachi, it would be imams who request help from some community members or accompany the supervisors themselves. Therefore, in the engagement of this international private sector organisation with the mosque, the mosque facilitated access, bridged the cultural gap and contributed to their work by helping establish their initial contact with remote communities.

Similar to the help of mosques and imams given to the supervisors of Mott MacDonald, some other private sector consultants also engaged with the mosque and the imam during the reconstruction and rehabilitation in the aftermath of the earthquake. Socio-engineering consultants were acting as consultants under the Earthquake Emergency Assistance Programme for the reconstruction of schools in the earthquake-affected areas in an Asian Development Bank-funded project. The Earthquake Emergency Assistance Programme had a soft component for the establishment and capacity building of the School Management Committees along with the reconstruction of the physical infrastructure of schools. The role of the local religious leader was particularly mentioned and suggested as a member of any typical School Management Committee in the School Management Committee handbook (Socio-engineering Consultants, 2006). This coincided with the inclusion of the religious leaders among key members of the Union Council Disaster Management Committee established under the CBDRM programme concurrently run by ERRA and funded by the World Bank.

In general, private sector entities coming from outside the affected communities connected with the mosque to negotiate their initial contact with affected communities. Depending upon the nature of the required business and its timing after the earthquake, characteristics of this interaction with the mosque varied among different private sector entities.

The next section summarises the overall role of mosques and imams in separate locations based on the fieldwork research.

4.6 The Overall Role of the Mosque in Disaster Management

The role of the mosque in disaster management differed from place to place depending on the particular characteristics of a community. The roles of mosques and imams in Banda-1 and -2 were similar but different from the role of the mosque and imam in Banda-3. Table 4.2 offers summary findings of this comparison of the roles of mosques and imams in disaster management in their communities. Based on fieldwork, this comparison presents similarities and differences in the roles of these

Table 4.2 Comparison of the roles of mosques and imams in disaster response, relief, recovery, reconstruction and rehabilitation in their communities

Role Dynamics	*Banda-1, -2 mosques*	*Banda-3 mosque*
Cultural The bridging cultural gap between different disaster management actors and the local community	All mosques served as an entry door for civil society, the private sector and government organisations coming to help and work with the earthquake-affected communities. Mosques and imams functioned as facilitators and bridges to introduce and build rapport between development partners and host communities. This was essential to win the support of communities and to avoid friction and conflict between outside organisation and local communities	
Psychosocial Influence on disaster risk perception and attitude towards disaster preparedness	All imams had a critical role in shaping disaster risk perceptions through the institution of the mosque. The mechanism for influence included sharing views in public, delivering open talks (Banda-1, -2 and -3) and Friday speeches (Banda-3). The earthquake was strongly interpreted as an 'act of God' through references from religious narratives. This interpretation, although useful for creating resilience, hindered practical steps to adherence to safety measure such as seismic safe building codes	
Spiritual well-being, healing and resilience	Mosques provided religious services such as imams leading prayers and teaching the Holy Quran to children. All communities noted that they were advised by mosques to stay calm and resilient, help each other, and refrain from creating disorder and stealing since it was a testing time for them. All communities acknowledged the psychosocial and spiritual support provided by their imams through private and public counselling. This healing directly contributed to the resilience among communities	
Economic Sharing of market information	All mosques provided a social space to make collective economic decisions such as harvesting crops and sharing information about the availability of employment opportunities and seasonal labour in local markets in the aftermath of the earthquake	

Role Dynamics	*Banda-1, -2 mosques*	*Banda-3 mosque*
Social		
Women—exclusion and inclusion	In all locations in the case study, women were not seen to be using mosques as physical places. Mosques did not have a provision for women to be able to pray with the congregation, although women desired this. Women said they would use mosques as emergency shelters in case of a future earthquake. The mosque as an institution supported women's involvement in culturally appropriate development activities such as stitching and midwifery training	
Bridging information gaps	All mosques helped bridge information gaps among different actors, including civil society, the private sector and the government. All mosques were frequently utilised by all actors to make public announcements through loudspeakers and word of mouth. Women received information through direct announcements from the mosque and men and children	
Community welfare	Imams did not engage in activities outside mosques as it was deemed inappropriate by the communities they served	In addition to religious services, the imam of the mosque engaged in acts of general welfare such as voluntary teaching in the local school and motivating the community to help each other after the earthquake

(continued)

Table 4.2 (continued)

Role Dynamics	*Banda-1, -2 mosques*	*Banda-3 mosque*
Networking with other actors such as civil society, private sector and government organisations and identification of the poor	Imams responded to civil society, private sector and government organisations during the disaster management cycle when contacted but did not pro-actively engage by themselves. Imams provided some advice to development organisations looking for the marginalised (such as widows and elderly) and the poor to support them in cash and kind	The imam actively engaged with all actors during the disaster management cycle. In one case, he took the leadership role by opting to be the head of the Parent-Teacher Council of the school. The mosque space was used for boys' school and community meetings. The imam was proactive in identifying the marginalised and the poor after the earthquake and connecting them with development organisations providing support in cash and kind
Influence of a community's perception of the role of the mosque and the imam	Communities did not like the use of mosques for non-spiritual purposes such as public meetings, nor imams performing general welfare functions	The community was supportive of the role of the mosque and the imam in non-religious activities such as conducting community meetings in the mosque and the imam's involvement with development organisations
Political Interaction with public representatives	Mosques and imams had no interaction with political figures	The mosque, through its imam, had a constructive collaboration with local political figures during the recovery, reconstruction, and rehabilitation phases

mosques and imams in five dimensions, namely cultural, psychosocial, economic, social and political.

On the whole, the institution of the mosque played a distinctive role in disaster management by affecting the social, cultural, psychosocial, spiritual, political and economic dimensions of the lives of the earthquake-affected communities in rural areas. Socially and culturally, the mosque served as an entry door, a bridge across cultural gaps and facilitated access to communities for private, government, local, national and international organisations during the earthquake response and relief. Unlike schools, hospitals and government organisations, which stopped functioning due to the destruction of their infrastructure, the institution of the mosque proved to be indestructible. The mosque functioned well beyond the limits of men and material, brick and mortar. The mosque edifice was demolished, but the mosque's institution survived.. It continued to function, surviving community men gathered to play on the rubble or in the open, and the mosque served as the collection point of the community even in the hardest of times. Psychosocially and spiritually, the imams provided psychosocial and spiritual healing to members of the Muslim faith and influenced their attitude to disaster preparedness. Religious interpretations of the earthquake caused communities to turn to God and to increase meditation and prayer. On the one hand, the psychosocial and spiritual healing by mosques fostered the resilience of the earthquake-affected people. On the other hand, religious interpretations of the earthquake promoted fatalistic tendencies, which negatively affected communities' attentiveness to DRR measures suggested by the government. Politically, the government and private sector accessed communities to convey their policies and strategies to build connections and win support for recovery and reconstruction projects. In financial terms, the affected communities exchanged livelihood-related information and coordinated their income-generating activities using the institution of the mosque.

4.7 Other Roles of the Mosque

All sectors of society, both public and private, that were approached during the fieldwork, agreed on the massive, untapped potential of the mosque for much-needed social transformations, ranging from individual reforms to changes in collective behaviour. The Director-General of Provincial Reconstruction and Rehabilitation Authority referred to the mosque as "the epicentre of society" (Director-General, Provincial Reconstruction and Rehabilitation Authority, Abbottabad, 05/2009). Below are some of the potential roles of the mosque and other community-based religious institutions identified by research participants and observed during the fieldwork.[14]

4.7.1 Social and Economic Roles of the Mosque

One key informant, a teacher, described her idea of the potential role of the mosque as a functional community institution (Female Key Informant, 36, Banda-1, 07/2009). She suggested community engagement during any future earthquake. She proposed that an announcement be made using the mosque loudspeaker for the people to leave their homes for open and safe places during an earthquake. Once gathered at a safe place, they should be advised to stay calm and patient until a plan is drawn up to tackle the situation. Consultations and deliberations should be held to find out ways and means to mitigate risks, save human lives and reduce losses. Meanwhile, the mosque should deliver the messages of endurance, patience and peace to reduce panic.

Mosques functioned as information centres for communities, providing updates on vital information including the arrival of relief items, development and private sector organisations and any reconstruction policies of the government. This potential use of the mosque becomes even more critical in contexts where there were only five televisions and two radios scattered through a community of 5,000 people (Banda-1). Therefore, the formal linking of the mosque in the disaster management cycle can provide further advantages in the smooth flow of information. Besides, rich and poor men in communities were seen talking unreservedly and social barriers diminished within the boundary of the mosque. Mosques would offer a suitable social space to allow community decision-making in a socially comfortable ambience.

Imams helped identify the poor and the needy in the aftermath of the 2005 earthquake, facilitating the process of linking the marginalised with

[14] Some parts of this section draw on my work during doctoral thesis titled "exploring the role of the mosque in dealing with disasters: A cases study of the 2005 earthquake in Pakistan" at Massey University, New Zealand available at http://hdl.handle.net/10179/4080.

development organisations. In addition, communities used prayer gatherings to communicate and decide among themselves about economic activities like the collective harvesting of crops.

The chairman of ERRA stated that the mosque had such an enormous potential that it could "revolutionise" society (Chairman, ERRA, Islamabad, 05/2009). He proposed that by raising the capacity of the imam and making him a resource person in the community, many development objectives could be achieved. He suggested that the imam should be included in disaster management and then perform at least the following three roles:

1. undergo first aid training
2. be a teacher of the school in the mosque after *Fajr*[15] until *Zuhar*[16] and inculcate disaster preparedness values of safety in children
3. be trained in other locally relevant crafts to make him free from entire reliance on charity and use his local leadership capability for the common good. For example, imams in the Swat area could be taught beekeeping.

He further explained that ERRA trained 250,000 people as masons/artisans in makeshift tents and shops, so mosques could be a place to impart such capacity-building skills to communities (Chairman, ERRA, Islamabad, 05/2009). He referred to this scenario as "a win–win situation" for communities, public and private development organisations and the state. Along these lines, the mosque could promote other livelihood-strengthening activities such as providing a training centre for capacity-developing trades. However, the use of the physical space of the mosque for non-religious roles might face resistance from communities, as was found during this research (Banda-1 and -2).

Coincidentally, I saw the chairman's suggestion in action in Banda-1 in July of the same year. A community toilet was being constructed along with the mosque as a part of masonry training for the village people. When I spoke to the NGO official, he said that they were training the local people to raise their capacity in different skills. In this connection, they thought to build a toilet for the community along with the

[15] The morning prayer offered before sunrise.

[16] The mid-day prayer, which is prayed once the sun has crossed the meridian.

mosque as a demonstration of masonry for the village people who did not have toilets in general (Male villager, 22, Bara, 07/2009). He believed this would have dual benefits: it would raise the capacity of the village people and also earn goodwill for their organisation since it was the first community toilet for a population of 5,000 people (Banda-1).

4.7.2 *Educational, Health Support, Political Empowerment and Advocacy Roles*

Many potential roles for the mosque were also suggested by the Council of Islamic Ideology in its report to the government about social reforms in Pakistan (Islamic Ideology Council, 1993). Established in 1962, the Council of Islamic Ideology is a constitutional body whose mission, functions and rules of procedure were provided in Articles 228–231 of the 1973 constitution. The Council's main responsibility is to recommend to parliament and provincial assemblies ways and means to enable Muslims of the country to order their individual and collective lives by the rules articulated in the Holy Quran and Hadith. The Director-General, during his interview, stated that the Council's recommendations are sent to the government, although the government is not bound to follow its recommendations (Director-General, Council of Islamic Ideology, Islamabad, 07/2009). The Director-General provided me with a 1993 report of the council.

The council reiterated in its report that the government had a great under-utilised resource in the form of a local social institution—the mosque. It cited the instances of multi-dimensional historical roles of the mosque benefitting communities. Most importantly, the report notes that the mosque was the best place to join, network and integrate the marginalised and the poorest of the poor into the main stratum of society with the welfare provisions of the state.

This study illustrates that these roles are inherently linked with disaster management as shown in Table 4.3. The mosque could be used for educational, health and empowerment purposes, all of which would directly contribute to effective disaster management. However, it may be noted that establishing such roles for the mosque may require a profound change in societal cultural norms, including acceptance of a more active role of women in the mosque.

Table 4.3 Potential roles of the mosque as explored during the fieldwork and suggested by the Council of Islamic Ideology

Roles	*Explanation*	*Impact on disaster management and development*
Educational	Arrangement for using mosques as primary schools after a disaster	Disasters often destroy educational facilities; yet getting children back into regular routines such as schooling, helps them recover after a disaster. Holding classes in the mosque makes them accessible to the poorest of the poor—this would raise the capability of vulnerable communities and contribute to effective disaster management. Providing classes in the mosque also presents an opportunity to impart disaster awareness and preparedness training to the children of the poor who otherwise might miss out. Such training could engage imams

(continued)

Table 4.3 (continued)

Roles	*Explanation*	*Impact on disaster management and development*
Basic Health Units, distribution of Zakat and welfare funds	Arrangement of Basic Health Unit along with mosques so that hassle-free health services and provision of Zakat and charity to the marginalised section of the society	Considering the case of remote village areas, people desperately need a health service close by. The mosque could improve effective disaster management by serving as a first-aid centre during response and relief phases of any future earthquake or other natural hazards. Imams, members of the mosque committee, and some local women require first aid training. In addition, the mosque committee can identify the needy and the poor at the local level for the distribution of Zakat and charity
Communication and advocacy	Requiring local and higher government. officials to consult with imams and have women-inclusive public hearings in or outside the mosque	This would directly contribute to advocacy, campaigning and empowerment of vulnerable communities including women by having increased chances of communication and information directly from public officials via a trusted channel such as the mosque. This would be equally useful for pro-poor development and effective disaster management

Roles	*Explanation*	*Impact on disaster management and development*
Information and coordination centre	Recognising the mosque as a key local institution to coordinate and organise disaster relief and DRR activities at the local level	Outside organisations can use the mosque to organise their response and relief activities such as identifying vulnerable households and distributing goods. Connecting the mosque with the early warning system and other support networks would further strengthen capacities and boost recovery processe
Psychosocial healing and spiritual well-being	Role of mosques and imams in promoting psychosocial healing and spiritual well-being at the local level	By training imams, mosques may be used more effectively for providing psychosocial healing and spiritual well-being during disaster recovery and rehabilitation by increasing resilience among communities through religious interpretation of disasters. In addition, this approach may also result in the maintenance of peace by reduction of instances of stealing and theft in the aftermath of disasters

(continued)

Table 4.3 (continued)

Roles	*Explanation*	*Impact on disaster management and development*
Other development functions	Establishment of mosque committees comprising local men and women who can engage in development and post-disaster activities	The mosque committee could devise culturally appropriate interventions through the following measures: 1. Advocacy for provision, repair and maintenance of civic facilities 2. Helping needy, widows and orphans 3. Establishment of communal library and literacy centre for illiterate in mosques 4. Educated individuals to devote some of their time to educate illiterate people of the community 5. Promotion of sectarian harmony

4.7.3 Role of Religious Institutions Helping Their Followers and Followers of Other Religions

Another potential benefit of engagement with religious institutions is the help the followers of one religion could offer to the followers of another religion. One such example was of a church that I had a chance to observe during my fieldwork. This church was accommodating some Muslim families along with Christian families who were displaced because of the military operation in Swat, providing them with food, shelter, security and other necessities of life. When I asked the bishop of the church about accommodating Muslim families, he responded "these are Christian teachings – Christians, Hindus and Muslims all pray, we are doing it just on a humanitarian basis without any hidden agenda" (Bishop of Mardan Church, 06/2009).

A range of potential roles for community-based religious institutions as identified by research participants shows huge under-utilised opportunities in all phases of the disaster management cycle. Distinctively supported by communities and located at a local level, these institutions could be used to save lives in disaster response and relief, coordinate and organise communities in disaster recovery, reconstruction and rehabilitation and prepare disaster-resilient communities by spreading awareness.

4.8 Conclusion

This chapter addressed the two research questions: firstly, it explored the role of the mosque in the disaster management cycle about other key players in the aftermath of the 2005 earthquake; secondly, it identified the potential future of the mosque. The findings have shown that the role of mosques varied from place to place. The mosque, community and the imam of Banda-3 actively engaged in the whole spectrum of the disaster management cycle in the aftermath of the earthquake. However, imams and mosques of Banda-1 and -2, although contributing to the response, relief, recovery, reconstruction and rehabilitation phases, were not as active as the mosque and the imam of Banda-3.

The findings have indicated that the community's view of the role of the mosque and its imam determined the limits of their engagement. The function of the mosque was democratic. It was mainly because of respect for the communities' choice in Banda-1 and -2 that their imams should not engage with development organisations directly but instead provide them with information to identify and assist the neediest and the poor in their villages. These rural communities did not like their imams to engage in secular acts (as they perceived them) besides worship.

Although there were differences in the role of the mosque in separate locations, all mosques served distinctly from other key actors in disaster management. They served as facilitators and enablers for outside disaster management actors, as a physical space and forum for coordinating response and relief, as a socially integrating force for the earthquake-affected, as a recruiting ground for volunteers and as an initial contact point during the response and relief phase in the aftermath of the earthquake. During the recovery, reconstruction and rehabilitation phase of the earthquake, mosques strengthened livelihoods and offered psychosocial support and spiritual healing to the local people.

The private sector appeared as both patron (where the concerned business people were affiliated with the mosque as members of the Muslim faith) and beneficiary of the institution of the mosque. At the local level, business individuals provided the mosque with an alternate space in the immediate aftermath of the earthquake as members of the Muslim faith (in Banda-1). In addition, the mosque served as a place through which to establish initial contact and connect the private sector with rural communities. The mosque bridged the cultural gap, was an expediter and facilitated contact with communities. The mosque, through the imam, negotiated the religious and cultural boundaries of purdah in remote communities.

The research has shown that the mosque as a physical place is not promising for women's engagement in disaster management in its current form. However, the support of the mosque is critical for women's engagement in disaster management activities outside the mosque. Women had strong affiliations with the mosque, although they did not physically visit mosques often. Due to cultural norms in the case-study area, women observed purdah and were not seen regularly in markets. Notwithstanding this, the reasons given by the village men for the absence of women from the mosque were also related to economic factors, since they were generally poor and could not free up enough resources to build separate provision for women in the mosque. In addition, there were some religious restrictions on women's visits to mosques, including staying off-site during menstrual periods.

However, women were connected with the mosque through the wife of the imam, their children learning the Holy Quran, their men offering prayers and them listening to the weekly Friday sermon. They had strong spiritual connections with the mosque. They considered the mosque as their survival strategy and maintained they would take shelter there along with children during a future earthquake. Since communities rebuilt their mosques without any respect for seismic safety codes, this poses a significant risk to the lives of these women and children.

All mosques included in the fieldwork influenced the perception of disaster risk to communities. Imams had parallel interpretations of the earthquake—one scientific and the other religious. The religious interpretation considered the earthquake more as a sign of a punishment or a test from God than a movement of tectonic plates. All communities ascribed this perception to messages from mosques through imams. On the one hand, the religious interpretation proved to be a source of psychosocial support in several ways, such as inspiring people to help each other and promoting resilience. On the other hand, due to the religious interpretation, communities did not take important practical steps to save themselves from future building hazards, for example, they avoided government-prescribed seismic-safe housing codes. However, communities were not entirely fatalistic as they actively participated in the CBDRM programme and had a powerful desire for more such disaster-preparedness initiatives by the government.

This research has exposed a gap in understanding between the standpoints of policymakers and those of "on the ground" communities. The former insists on adopting a unified approach without due respect for communities' views of hazard and vulnerability. Such standpoints (of ERRA) ignore the fact that there is a strong religious discourse on the causes of disasters in Muslim's scriptures, the Holy Quran and Hadith. This religious discourse, considering earthquakes either a punishment or a test from God, is systematically reinforced in the minds of men and women of communities through regular messages from mosques through imams.

Therefore, in a faith-based society where people's religious beliefs shape their worldviews of disasters, a standardised top-down disaster management policy and institutional structure oblivious of the inclusion of the community's primary institution may not deliver its anticipated outcomes of a safer life. In addition, the study has revealed that although there is an acknowledgement of the huge potential for the role of the mosque at the policy level, unexamined concerns (such as the mosques' conservative approach and sectarianism) remain and impede a meaningful engagement of the government with the institution of the mosque in all phases of the disaster management cycle.

At the grassroots level, this research has discovered that the front-line functionaries of the government (ERRA CBDRM master trainers) and international organisations (such as UN-Habitat, USAID and Mott MacDonald) appreciate the presence of the mosque as a useful institution to get their job done. They involve mosques and imams, understanding the importance of their role and influence on public opinion. They endeavour to reduce the gap between the standardised one-size-fits-all disaster management policy and local socio-cultural context by negotiating with mosques and imams and finding an innovative approach. By incorporating religious texts about the occurrences of earthquakes (and other natural hazards), communities were more prepared to participate in earthquake recovery and reconstruction. Imams were involved since they were the credible mouthpieces of religious teachings and popular opinion-makers in communities. The involvement of imams engendered community trust and ownership in these organisations' recovery and reconstruction efforts.

The research has exhibited that all key actors in disaster management acknowledge the huge untapped potential of this religious-cum-social institution. This is despite the mosque's sub-optimal role in the aftermath of the 2005 earthquake in its current form. Overall, the mosque played a distinctive role in contributing to different phases of the disaster management cycle in the social, cultural, psychological, spiritual, political and economic dimensions of communities in the aftermath of the 2005 earthquake. This study has documented the prominent role of the mosque in its capacity as a community institution. Whether an imam performed an active or a passive role in the response, relief, recovery, reconstruction and rehabilitation phases, he was an important stakeholder whose inclusion or exclusion could have considerable implications for the success of disaster management strategies. Does the question remain how best to engage him?

This chapter has indicated the need to search for an approach to disaster management that could accept religious interpretation as a valid social explanation along with the mainstream scientific understanding of natural hazards. There is a need to understand the reality of the earthquake through the eyes of the affected population. There may be a possibility of benefitting from a morally contextualised religious interpretation, which inspires people to help each other, provides psychosocial support through the creation of the shared meaning of a disaster and promotes resilience. This approach could provide a synergy by combining religious and scientific explanations of disaster while curtailing the negative aspects of fatalistic inclinations.

Chapter 5 will further elaborate on this aspect and other conclusions from this chapter by comparing them with the literature. It will also show how this research informs academic scholarship and the concept and practice of disaster management for its further application in similar situations including COVID-19.

References

Earthquake Reconstruction and Rehabilitation Authority. (2011). *Annual review 2009–2010*. Earthquake Reconstruction and Rehabilitation Authority (ERRA).

Islamic Ideology Council. (1993). *Social reformations* (in Urdu). Department of Computer and Publication.

Socio-engineering Consultants. (2006). *Capacity building handbook for school management committee: Earthquake emergency assistance programme (EAAP)*. Socio-engineering Consultants.

Times of India. (2009). 1.6 million Pakistani refugees return home: UN. *Times of India*. http://articles.timesofindia.indiatimes.com/2009-08-22/pakistan/28191405_1_refugees-return-home

CHAPTER 5

Opportunities and Challenges of Engagement with the Mosque as a Community-Based Religious Institution

5.1 Introduction

This book aims to examine the role of the mosque as a community-based religious institution in disaster situations. To achieve this aim, two research questions were addressed. The first examined the role of the mosque in relation to the key actors from the state, civil society and the private sector during the response, relief, recovery, reconstruction and rehabilitation in the aftermath of the 2005 earthquake in Pakistan. The second explored the potential roles of the mosque in similar situations in the future. A case study of the district of Mansehra in the Khyber Pakhtunkhwa province of Pakistan was presented. Using qualitative research methods, a broad range of actors in disaster management, such as earthquake-affected communities, imams, representatives of civil society organisations, people from the private sector and government organisations, shed light on the above questions.

This chapter discusses the findings of this book concerning the literature on disaster management about the role of community-based religious institutions in the context of post-development theory. This chapter concludes and illustrates the distinct contribution of this book to the body of knowledge. Before this research, the role of the mosque had mostly been hidden, undocumented, underestimated and unacknowledged.

A. R. Cheema, *The Role of Mosque in Building Resilient Communities*, Islam and Global Studies,
https://doi.org/10.1007/978-981-16-7600-0_5

This chapter also suggests new dimensions for exploration to further improve our understanding of the role of community-based religious institutions in disaster management both at conceptual and practical levels.

5.2 Contribution to Knowledge About the Role of Religious Institutions

Overall, this book corroborates the finding that community-based religious institutions play a key role in developing cohesion and building social safety which is critical for people's recovery and rehabilitation in the wake of disasters. Some of the particular roles identified during this research are discussed in the following sub-sections.[1]

5.2.1 The Centrality of Religious Institutions to Communities

The mosque, as a vital community hub and gathering place for men to offer prayers five times a day from dawn until night, served as a logical point for interaction, negotiation and communication for the government, civil society and the private sector. One of the extraordinary attributes of the institution of the mosque in the wake of the 2005 earthquake was its undisrupted functioning despite the deaths of its patrons and the destruction of its brick and mortar building. The uniting force of religion saw its expression in communities' affiliation with the mosque, attracting congregations of men for prayers, in the open or an alternate place, even when the mosque infrastructure was partially or fully destroyed. These congregations created a social space which promoted a spirit of mutual help and sacrifice among communities during times of extreme misery and destruction in the aftermath of the earthquake. This is similar to the role of religious institutions found elsewhere and the roles they play in disasters including the prevalent pandemic—COVID-19 (Al-Astewani, 2021; D. K. Chester et al., 2008).

The mosque reflected the two core characteristics—community ownership and trust—essential to make any development project a

[1] Some parts of this section draw on my work during doctoral thesis titled "exploring the role of the mosque in dealing with disasters: A cases study of the 2005 earthquake in Pakistan" at Massey University, New Zealand available at http://hdl.handle.net/10179/4080.

success (Chambers, 1997, 2005; Sen, 1999). The communities studied made the land available and built their mosque by themselves. They appointed their imams themselves. They used the mosque and sustained its expenses including payment to the imam. The mosque was far more than a religious place of worship for them; it was a spiritual defence and part of their survival strategy to safeguard them from natural hazards. These findings are similar to McGregor (2010) who found that Muslims in Aceh wanted to build their "meunasah" (local village mosque) before their own houses during the post-tsunami reconstruction. This was because of the critical significance of the 'meunasah' as a centre for psychosocial support and spiritual healing of communities, essential for their recovery. It was not only a prayer place but also a forum for sharing happiness (e.g. conducting marriages) and sadness (e.g. interpreting tsunami destruction) in their day-to-day lives. Therefore, this book contributes to the scarce literature in development and disaster studies by highlighting the importance of community-based religious institutions. The findings also resonate with the work of others (Bruinessen, 2019; Feener, 2013; Fountain et al., 2015) who emphasise the dynamic role of religious institutions and communities in varying contexts and the extraordinary capacity of these institutions to adapt, exist and thrive.

This exploration of the role of the mosque has elucidated that religion (through its community institutions), which greatly helped them to interpret meanings of death and destruction after the 2005 earthquake, provided a cohesive force for communities, and promoted resilience. Religion as a socially integrating force and faith as a form of social capital can play a role in community development. This particularly occurs at tough times such as in the aftermath of natural hazards when people refer to their divine belief more than usual to find support and pacify the sense of loss. The prayer gatherings created social-cum-faith capital (Candland, 2000; Sheikhi et al., 2020) which promoted a spirit of mutual help and sacrifice in communities during times of extreme misery and destruction in the aftermath of the earthquake. Religion was not only a defining feature of communities' worldviews, it also deeply contributed to the social, cultural, economic and political patterns of their approach to life.

5.2.2 *The Need for Partnership Between Outside Organisations and Community-Based Religious Institutions in the Face of Disasters*

Mosques were engaged to varying levels during the recovery, reconstruction and rehabilitation phase of the earthquake, but their contribution was often hidden. This varying level of engagement did not occur due to a conscious choice (such as the result of a policy towards engagement with religious institutions, except in the case of USAID) but because of their position as a key local community institution. The frontline staff of the state, civil society, private and public sector organisations engaged with mosques for practical reasons of reaching communities and gaining legitimacy for their work. This decision was given the reality on the ground—the appropriate way to approach communities. The action of the frontline staff appeared consistent with Chambers' advice (Chambers, 1984, 2002; Hirschmann, 2003): to recognise who mattered the most in reality. Focus group discussions with communities confirmed that the field staff of different national and international organisations (e.g. ERRA and UN-Habitat) visited mosques to relate to and interact with communities. These staff involved imams to explain to communities the importance of seismic safety codes and of learning about rescue techniques by narrating to them the Prophet Noah's story from the Holy Quran. The findings show that the social conduct of the mosque is strongly influenced by the organisation of the people (the mosque committee and communities) and influential individuals (the imam). It supports the case for emphasising the role of people's organisations and core individuals to achieve development outcomes that are significant and community centred. This underpins the contention that "institutions are not self-generating or self-sustaining and they achieve little on their own" Leftwich & Sen (2010). The imam's support or opposition to a development activity at the local level may have a significant effect on its results. As found by Atallah, Khan and Malkawi, "Imams are more capable of reaching the public than water specialists". The findings demonstrate that the mosque has a key role in bridging the gap between outside and inside stakeholders and thus contributes to improved understanding between the two, resulting in the reduction of disaster risk and sustainable development outcomes. This role of local institutions in bridging the gap is demonstrated in other work including in the ongoing endeavours to overcome COVID-19 (Al-Astewani, 2021; Feener & Fountain, 2018; Heijmans et al., 2009; Wisner, 1998).

In this context, the study identifies the critical role of the mosque for disaster policy and practice. The mosque as a community institution is a key actor in disaster management, and the imam is a key individual who influences how the community perceives and deals with disaster risk. Thus, this study argues the case for engagement with community-based religious institutions and their leaders in support of an earlier finding that faith leaders are in a distinctive and critical community leadership position and they influence the achievement of public policy on DRR, nutrition and health (Gaillard & Texier, 2010; M. Clarke, 2013; *The Dawn*, 2020; Vyborny, 2020).

All seven imams and one Christian bishop interviewed during the fieldwork research showed the willingness to work with the state for the religiously noble cause of saving people's lives from the vagaries of natural hazards.

This book's results are consistent with earlier views that religious institutions were taken on board if they were perceived to be supportive of development activities but ignored if perceived otherwise (Bano & Nair, 2007). Although a majority of public sector top officials acknowledged the exceptional potential of the mosque in the future, they did not formally approve interaction with the mosque during the disaster management cycle in the aftermath of the earthquake. This was because the institution of the mosque was perceived either as unrelated or in opposition to disaster management in its current form (even though the religious leaders were willing to support any effort towards disaster management). There could have been practical reasons, such as the pressing needs for food and shelter in the immediate aftermath of a disaster, which impeded new partnerships. However, this book emphasises the importance of more research to explore social viewpoints, in particular those of religious institutions, to reduce this gap not only in the disaster studies literature but in public health, economics and management fields (A. Oliver-Smith & Hoffman, 1999; David K. Chester & Duncan, 2010; DeFranza et al., 2020).

Given this, the research supports the emerging stance that developing partnerships between international organisations and local community-based religious institutions are important to effectively increase local ownership, win the trust and utilise local knowledge, organisations and resources (Benthall, 2017; McClure, 2017; Reale, 2010; Suri, 2018; Vyborny, 2020). In the post-2005 earthquake context, the study has shown one example of an international organisation (USAID) that worked successfully with imams and mosques during the earthquake recovery, reconstruction and rehabilitation phase as a matter of policy. These involvements with the mosque led to a solution where everyone benefits for communities and the donor. There was strong community support and involvement, for example, for shifting a girls' school to a boys' school building and the boys' school to the mosque in Banda-3.

5.2.3 *Religious Institutions and Gender Issues*

This research also raises gender issues regarding the future role of the mosque in the disaster management cycle. It agrees with earlier findings that religiosity strongly influences attitudes towards gender equality and the importance of taking a complex religious approach to study religion and sex-related outcomes (De Cordier, 2010; Krull et al., 2021; Seguino, 2011).

Women were not seen inside mosques. They could not join prayer congregations although they had a desire to do so. They did not form part of the decision making and consultation meetings about disaster management held inside mosques, although there were other channels, such as the wife of the imam and the men of their families, through which they stayed informed and involved. Due to cultural reasons (such as strong purdah custom), religious reasons (such as limitations of physical purity preventing entry to the mosque) and financial reasons (lack of resources to build a separate provision for women), the mosque as a site does not seem to be a promising avenue for the involvement of women in disaster management in its current form. Despite these current limitations, this research identifies that religious influence is not unavoidably negative and the mosque as a forum can be utilised to further gender participation in disaster management. The female research participants who were literate had become so because of religious education that was accessible, free of cost, supported by imams and provided by seminaries attached to mosques. These women, in addition to learning basic numeracy skills, learnt how to read and write. The imams promoted socially appropriate income-generating and welfare activities for women such as sewing and mid-wife training, respectively.

The need for gender-sensitive approaches during all phases of the disaster cycle is increasingly emphasised (Horton, 2012; Oktari et al., 2021). How can gender-sensitive disaster management practices thus be promoted through the mosque? This research provides evidence of women's involvement in disaster management, as was the case in the Community-based Disaster Risk Management Programme (CBDRM), where their participation was secured through the support of imams. If appropriately negotiated and engaged, mosques, in an institutional capacity, could be used to promote women's participation in disaster management. Most importantly, in an institutional capacity, the backing or opposition from the mosque can make a significant difference to the success or failure of disaster management policy and the empowerment of women in particular. This finding corroborates Wisner's (2010) view that even a little support from religious leaders would make a big difference to the secular efforts of DRR. In cultural settings where women are not visible in public places, the mosque performs the role of a "gatekeeper" (Crang & Cook, 2007) negotiating to facilitate the initial contact of outside actors (from the state, civil society and private sector) with communities. This occurred when imams assisted the staff of an international private sector firm (Mott MacDonald International UK) to carry out third party validation of housing reconstruction after the 2005 earthquake. The staff used the mosque to introduce them to the community, to stay in overnight. They used the services of imams to talk to women whose husbands were absent. In such an environment, the mosque is a critical community institution that can foster social acceptance for women to participate and contribute towards effective disaster management. These findings are in line with the growing realisation among international organisations (such as DFID, ICRC and UNICEF) of the value of engagement with international faith-based organisations based on their local grassroots connections including religious institutions (Bonney & Hussain, 2001; IRRC, 2005; UNICEF, 2017, 2020).

Another example of the mosque and imams supporting the fight against poverty with a civil society organisation (Akhuwat, 2021) was found during the fieldwork and expanding its services over time. This organisation, Akhuwat, used the mosque as its office to hold gatherings, saving money on its operational costs and the mosque environment to employ a moral responsibility for returning interest-free loans offered to the poor. The distinctive advantage of this approach, having the mosque and other religious institutions such as churches and temples as a recognised community institution in disaster and development contexts, could be its in-built cultural appropriateness, local ownership, broad participation of communities and a therefore greater chance of success in achieving the objectives of development and disaster management strategies and policies.

5.2.4 *Moving Beyond Political Sensitivities Associated with the Mosque*

In the political context, the study questions simplistic pejorative labels attached to the mosque, such as being a centre for preaching radical views in society and, therefore, not a safe partner in disaster management. Such labels belittle and fail to recognise the distinct position of the mosque as a central community institution (Abdel-Hady, 2010; Haque, 2009) drawing attention to the use of the mosque and its social conduct rather than the mosque itself. On this point, it presents a case for engagement rather than disengagement with the mosque as shown by some organisations and Muslim-minority western countries such as the USA (Jamal, 2005). As a community-based religious institution that is embedded in the everyday socio-cultural life of people, the mosque continues to exist, serve and enjoy its revered position in society. In this way, people will not change their priorities and abandon their affiliation with the mosque just because it has become contentious for the state and any other multi-lateral development or humanitarian role-player. The findings show that all the seven imams and one Christian bishop included in this research showed readiness to engage with the state for the religiously virtuous cause of saving human beings in the face of disaster. Therefore, this book argues that local, regional, national and international level disaster management institutions should acknowledge and engage with the mosque, and utilise its grassroots position for effective and efficient disaster management and to reduce the loss of human lives and assets.

This book adds to a growing understanding that it is difficult to ignore the increasing role of religious institutions where religion is one of the defining features and custodians of the social and cultural values and norms of communities (Atia, 2013; DeFranza et al., 2020; Feener & Scheer, 2018; ICRC, 2020). Disasters give rise to a situation where people from different parts of the world, quite unfamiliar with each other, come in contact to save lives, provide necessities and rebuild homes during the disaster management cycle. International civil society and private sector organisations have concerns about how to maintain neutrality, stay safe, ensure respect for local sensitivities and win the necessary support of communities to carry out their job. The findings of the fieldwork show that all the five mosques have been entry points to access communities during response and relief. One mosque (in Banda-3) even facilitated public and private sector activities during the recovery, reconstruction and rehabilitation phase. In cases where mosques and imams could not actively provide this support, this was not due to any inherent ideological conflict with disaster management activities on their behalf, but rather because of communities' reluctance to see their imams involved in apparently non-worship activities. In this context, the research draws attention to the fact that not all mosques and imams were "equal" with regards to their involvement in disaster management activities. There was diversity in how communities related to their mosques and consequently the roles that mosques played in post-disaster settings. Therefore, it is even more critical to work with this key community institution in both pre-and post-disaster times.

5.2.5 The Influence of Religiosity on Disaster Risk Perceptions

In the psychosocial and spiritual context, the conceptual framework developed in Chapter 2 emphasised the inclusion of the mosque because it significantly shapes people's attitudes and perception towards the physical and social environment they live in; and it is especially significant to their attitudes and actions in the face of disaster risk. The results are consistent with findings that disaster risk perception is fundamental to people's responsiveness, preparedness and ultimately to the promotion of disaster prevention culture (Anthony Oliver-Smith, 2004; Cannon et al., 2003). On the one hand, it confirms the findings of earlier studies such as Chester et al. (2008), who found that perceptions of disaster risk are strongly influenced by the religious views of faith-based communities. On the other hand, none of the seven imams interviewed during this study ascribed the earthquake to institutional sinfulness, such as the failure of the government to fulfil its responsibilities to minimise hazard risk before the earthquake. This finding differs from the growing trend among some of the Christian organisations and leaders who hold institutions responsible for disaster losses, as noted by Chester and Duncan (2010). However, when involved in training and engaged by UN-Habitat, Imams adopted physical interpretations of the earthquake and persuaded communities to build seismic-safe houses. Thus, there is the potential for them to take a broader perspective on the causes of disasters.

The book findings show that the communities had a parallel interpretation and understanding of disasters: religious and scientific. The international and national disaster management-related organisations base their policy structure on the scientific interpretation of disasters. People's (or popular) interpretations of disasters, which are religious more often than not in places where religion is a defining component of society, are neither considered nor expected to be accommodated by scientific interpretation. However, this book argues that ignorance of religious interpretations of disasters could lead to failure of compliance with disaster management policies. Therefore, the aim of securing lives from disaster losses through an integrative approach to disasters may not be achieved without understanding the religious context (where it matters) of disaster risk perception.

This research finds that the way the state or development organisations explain disasters is not always the same as how disaster-affected people perceive them and that mosques have an important role in influencing people's perceptions of disaster risk. People's perceptions were strongly influenced by the messages from the mosques. These perceptions had both negative and positive implications. Among negative repercussions was that the religious interpretation of the earthquake decreased people's willing compliance with disaster preparedness measures even in the aftermath of such a deadly calamity like the 2005 earthquake. In addition, communities did not fully adhere to seismic safety codes although they participated in other preparedness initiatives such as the CBDRM programme. This is different from the findings of Paradise (2005) who found that those who thought that the earthquake occurred because of the order of Allah were fatalistic. However, this book shows that research participants exhibited only some fatalistic tendencies, which is similar to Kennedy's (2001) findings which showed that about 7,000 Sicilians responded to state evacuation calls, and also held mass prayer gatherings to defend against the volcanic eruption.

Of the positive effects, resilience was enhanced during the disaster management cycle in the aftermath of the earthquake due to the contribution of the mosque. Prayer gatherings grew stronger and increased in size as more men returned to the mosques in the case study villages. The study confirms earlier findings that people believe they get closer to God during times of disaster (DeFranza et al., 2020; Gibson, 2006). Using this opportunity, the mosque through the imam offered a unique interpretation of the earthquake to the people. This interpretation was socially and culturally acceptable and it created resilience in communities to be able to cope with the trauma and shock, engendered hope and fostered psychosocial recovery. Communities were advised to repent, reconcile and mend their relationship with Allah and stay patient since the earthquake was either punishment of their sins or a trial. Other studies also acknowledge the contribution of religious institutions to increasing resilience in disaster-affected communities, e.g. see Laditka et al. (2010) for the support of churches after Hurricane Katrina) or otherwise mentally stressful situations, e.g. see Ni Raghallaigh and Gilligan (2010) for the contribution of churches in the lives of unaccompanied asylum seekers in Ireland). Keeping in view the positive and negative aspects of religious interpretations of disasters, it is only through meaningful engagement with mosques and imams that strategies are sought to curtail fatalistic tendencies and strengthen resilience among communities. This may help lead communities away from the path of reinforcement of vulnerability during the recovery and reconstruction phase in the aftermath of a disaster.

5.2.6 *The Role of the Mosque in Supporting Livelihood Recovery*

In the economic context, the findings confirm earlier results that the mosque plays a critical role as an information centre in underdeveloped rural communities at the local level (Belshaw et al., 2001; Rahmani, 2006; Régnier et al., 2008). The findings demonstrate that men's gatherings in the mosque led to livelihood related decisions such as collective watering and harvesting of crops in addition to an exchange of information regarding any employment opportunities in the area during the recovery, reconstruction and rehabilitation phase of the earthquake. The mosque was an intra-community information exchange centre; information came from outside and was spread from the mosque to the whole village through loudspeaker announcements. The mosque platform was also used by humanitarian organisations (such as World Vision) to seek the assistance of the imam in collecting information about the poor and the marginalised in a community. This information was to serve different purposes including disbursement of cash grants to the poor and devise programmes of income generation for them.

5.3 Conclusion

Overall, this book has contributed to the understanding of the role that mosques and other religious institutions can play in disaster management[2] It suggests that key actors from the state, civil society and private sector involved in disaster management need to understand the complex relationships (between people and their religious institutions), and their joint impact on the social environment. A collaborative approach with other actors based on mutual understanding might allow the mosque to realise its potential as a community-based religious institution. This attention and engagement might enable the mosque to perform the eight core functions of the two principles of the UNISDR strategy—directly "connecting" with and "convincing" 1.6 billion Muslims around the world how to work towards a safer future. The centrality of the mosque as a physical structure and a community institution means it is capable of connecting communities by coordinating, campaigning, advocating and informing about DRR activities. Equally, the mosque can convince local people by organising, promoting, encouraging and providing for DRR initiatives.

5.4 Improving Disaster Management in Pakistan

This section discusses key issues of disaster management in Pakistan as identified in Chapters 3 and 4 and provides some suggestions to address them. It also deliberates upon the significance of the role of the mosque in improving the effectiveness of the country's disaster management structures.

5.4.1 The Organisational Structure of Disaster Management

The findings show that the practice of the integrated approach to disaster management (Ishiwatari, 2021; UNISDR, 2005), as popularised by the United Nations since the 2000s, suffers from serious institutional shortcomings in the case of Pakistan. Despite new legislation (The

[2] Some parts of this section draw on my work during doctoral thesis titled "exploring the role of the mosque in dealing with disasters: A cases study of the 2005 earthquake in Pakistan" at Massey University, New Zealand available at http://hdl.handle.net/10179/4080.

Disaster Management Act 2010) and the framework (2007) to deal with disasters prepared with the technical assistance of the United Nations Development Programme (UNDP), patterns of horizontal and vertical functional overlap among government disaster-related organisations at federal, provincial and district levels have not faded away. In opposition to the prescriptions of the integrated approach emphasising local level capacity building, the Pakistan disaster management structure remains ad hoc and top-heavy with only marginal resource transfer to the local level in practice.

Investigation of 50 years of the development planning history of Pakistan has shown that the country's disaster management policies remained narrowly focussed on dealing with only one of kind of hazard (flooding) before the 2005 earthquake. The disaster management structure was overhauled after the 2005 earthquake in a bid to make it more effective according to the guidelines of the integrated approach. However, the study results illustrate that the addition of new actors increased complexity of the disaster management structure and created new institutional challenges such as friction and conflict of interest without solving the previous ones, including power and resource struggles. The pattern of erecting new disaster management structures without evaluating, upgrading, merging or dismantling the former continued even when the country was hit with one calamity (2005 earthquake), after the other (SWAT Internally Displaced Persons' crisis 2009 and then 2010 flooding). A fire-fighting attitude has dominated the disaster management practices despite the government's paper commitments to pursue an integrated approach to disasters in the new framework. These findings are similar to other studies on Pakistan identifying systematic patterns of vulnerability and narrowly focussed structural hazard approaches.

The government can get rid of ad-hocism by eradicating duplication of roles (e.g. the MoCC and NDMA) and implement resource transfer at the local level (to DDMAs and downwards to Union Councils) to benefit from the best practices advocated by the integrated approach to disasters in Pakistan. Meanwhile, the importance of community institutions such as the mosque becomes critical in saving lives and reducing disaster losses. Because government assistance may arrive later than expected and communities may have to fend for themselves for extended periods of time, it is vital to integrate DRR into communities' daily life through community-based religious institutions such as mosques. How disaster management structures could work more effectively with the mosque

The study has demonstrated that the mosque has played a multi-faceted and visible role in the response, relief, recovery, reconstruction and rehabilitation phases of the disaster management cycle in the aftermath of the 2005 earthquake. Nevertheless, this role of the mosque is far from realising its full potential. The readiness of all the seven imams to welcome any initiative soliciting their cooperation to save people's lives and properties is a critical finding for disaster management policymakers and practitioners. The findings support rare voices which argue that the great hidden potential of religious communities and their leaders could be tapped by engagement (Fountain et al., 2015; Séverine Deneulin & Rakodi, 2011; Wisner, 2010). The mosque is like an "entrance door" to communities and the imam is the "key" to this door. If the entrance door (the mosque) and the key (the imam) are engaged, involved and taken on board in disaster management, the ultimate goal of saving human lives and reducing disaster losses might be better achieved at the local level where disasters hit the hardest.

The suggestion of the chairman of ERRA interviewed during the research is worthy of attention: building the capacity of the imam through first aid and vocational training. Imams (equivalent to religious/faith leaders) need to be embedded and involved in disaster preparedness initiatives at the local level since they could become "agents of safety" (Rowel et al., 2011). In the aftermath of the 2005 earthquake, a strong suggestion has been made that the mosque and the Imam be engaged to announce imminent threat as part of an early warning mechanism at the local level (Dekens, 2007; K. Sudmeier-Rieux et al., 2008, 2011). Local religious leaders could undergo specialised disaster training that combines religious and scientific knowledge to create synergy. Moreover, such training need not be one-way. It could help key agencies involved in pre- and post-disaster situations better understand the nature of communities and the role of the mosque and the imam. However, the findings also show that a communities' perception of the role of an imam can also be a constraint on his social conduct. As found in Banda-1 and -2, despite their desire, the imams could not openly participate in relief and recovery activities since communities did not approve of their engagement in non-worship matters. Therefore, such training of imams may also build their capacity to educate and enlighten their communities about the social roles of the mosque. This book does not suggest that the mosque can solely fulfil all objectives of the Sendai Framework in its current form. Rather, it corroborates the emerging view that religion and spirituality be regarded as significant and that it is time to engage with community-based religious institutions as important stakeholders in disaster management including health pandemics of COVID-19 (Al-Astewani, 2021; BBC, 2021; DeFranza et al., 2020; Krull et al., 2021). It underscores the imperative for wider community training and education programmes where imams and disaster agencies work as partners to build a shared understanding of disaster risk and measures to reduce such risk.

This is the first in-depth case study of its kind examining the role of the mosque in disaster management in the context of Pakistan. The research findings support the literature arguing that the proper employment of the mosque, a hugely underutilised social institution, can help to address several social issues in Pakistan that the state cannot handle on its own (Islamic Ideology Council, 1993; Vyborny, 2020). The significance of the mosque as a public opinion-making institution of society was formally acknowledged by the government when the Ministry of Population trained and involved imams in birth control advocacy programmes (Ministry of Population Welfare, 2005; *the worlds' population*, 2020). The government can benefit from the Ministry of Population to look for more meaningful ways of engaging with mosques for disaster management. Although hesitant to engage with the mosque, the top policymakers of the government called it an immensely under-employed community resource that had ample potential to transform society and address social issues.

From where can a meaningful and sustainable partnership be set up, built with simple steps and maintained? COVID-19 has offered an unprecedented opportunity to the national and sub-national governments in Pakistan. Taking an inclusive and grassroots approach, imams/leaders of areas mosques/churches/temples or other community based religious institutions (or their representatives) should be made part of district/sub-district/town level development committees. Usually, these committees are in place in these setups and include representatives of government line departments and other development organisations working in the area. The meeting frequency is usually once a month. Unlike project-driven committees which appear and disappear, these development committees are there for good, to improve coordination, reduce duplication and build a collective endeavour place-based development agendas. This strategy should be rolled out in both urban and rural contexts. Just by including Imams in these committees can be a simple but strategic step to build trust and long-term partnership to build inclusive, equitable and sustainable societies and help achieve the Agenda 2030. There is room for a dedicated organisation, public or private, that could solely work for bridging gaps and building partnerships between mosques and disaster management and development organisations for building resilient communities.

5.5 At Crossroads: Religion, International Humanitarian Actors and Mitigating Impacts of Extreme Events on the Lives of People

This book has attempted to bridge the gap in the body of knowledge about the role of local community-based religious institutions in disaster management by highlighting the hidden, underrepresented but multi-faceted role of the mosque in the aftermath of the 2005 earthquake in Pakistan. It has created new knowledge about the role of community-based religious institutions, particularly the mosque, in the response, relief, recovery, reconstruction, rehabilitation and preparedness phases of a disaster. It has added to the body of theoretical knowledge on post-development theory by providing a new example of collaborating with communities through a community-based religious institution. In societies where religion is a defining feature of individual and collective identity, a powerful religious narrative and worldview are likely to permeate, if not prevail, in the socio-cultural environment. As disasters occur, community-based religious institutions such as mosques spread their views and interpretations among disaster-affected communities.

Given the rise in extreme events and to reduce the extent of consequent losses, empirical and systematic evidence is needed to reduce the gap in knowledge about the vital role of community-based religious institutions. Still, we know little about the countless ways in which community-based religious institutions, groups and movements work in practice in a highly dynamic, adaptive and contextualised manner in different parts of the world (Atia, 2013; Davis & Robinson, 2012; Feener & Scheer, 2018). A broad range of religious practices has survived centuries of an onslaught of various ideological and political challenges in Africa, Asia and North America. Disasters create a full range of opportunities for laying out a new landscape where new narratives can be built, old myths can be dismantled and social realities can be manufactured (Feener, 2013; Simpson, 2014).

Although 84% of the worlds' population affiliates to one of the many religious faiths (Pew Research Centre, 2012), religion is far from being mainstreamed in the work of large development and humanitarian organisations. At the best, most working arrangements stay ad hoc and transaction-oriented. In part, this is due to these organisations' lack of religious literacy.

This book argues that the makers of disaster management policy and structure should rethink their conservative stance of staying disengaged from religious institutions at both policy and practice levels. By documenting and analysing the role of the mosque in disaster management through an in-depth case study, it has disputed their wisdom of underestimating the invisible roles of religious institutions, by highlighting their ability to influence the social, economic and political processes that affect communities' response to hazards. It furthers the case for engagement and dialogue with community-based religious institutions including churches, mosques, synagogues and temples while acknowledging and valuing them as important stakeholders in disaster management. This book puts forward the case for adopting a new collaborative approach that acknowledges both the religious and scientific views in disaster management. Such an approach may enable us to better prepare and survive future disasters both materially and spiritually.

The policymakers both in public and private sectors, development practitioners and media personnel do not have a due understanding of religion. Instead of looking at the religious interpretations employed by diverse groups for their vested interests, religion itself has been seen as the major driver of conflicts. Due appreciation is lacking on both sides that religion and proclaimed secular humanitarianism remains intertwined with politics and development (Fountain et al., 2015; Gingerich et al., 2017). May it be poverty or other development concerns, deep religious undercurrents influence how development agendas take shape and translate into different policy actions.

Future research may suggest how to further benefit society by channelling the multi-faceted contribution of community-based religious institutions, such as their psychosocial and spiritual healing services, the trust and community ownership that they enjoy, and their ability to influence public perception.

* * *

References

Abdel-Hady, Z. M. (2010). *The masjid, yesterday and today*. The Centre for International and Regional Studies, Georgetown University School of Foreign Service.

Akhuwat. (2021). *Akhuwat's story* (Akhuwat, Ed.). Akhuwat. http://www.akhuwat.org.pk

Al-Astewani, A. (2021). To open or close? COVID-19, mosques and the role of religious authority within the British Muslim community: A socio-legal analysis. *Religions, 12*(1), 1–26. https://doi.org/10.3390/rel12010011

Atia, M. (2013). *Building a house in Heaven: Pious neoliberalism and Islamic charity in Egypt*. University of Minnesota Press.

Bano, M., & Nair, P. (2007). *Faith-based organizations in South Asia: Historical evolution, current status and nature of interaction with the state*. University of Birmingham.

BBC. (2021, January 21). *Birmingham mosque becomes UK's first to offer Covid vaccine*. https://www.bbc.com/news/uk-england-birmingham-55752056

Belshaw, D., Calderisi, R., & Sugden, C. (2001). *Faith in development: Partnership between the World Bank and the churches of Africa*. Regnum Books International.

Benthall, J. (2017). Charity. In F. Stein, S. Lazar, M. Candea, H. Diemberger, J. Robbinson, A. Sanchez, & R. Stasch (Eds.), *The Cambridge encyclopedia of anthropology* (pp. 1–15). Cambridge University Press. http://doi.org/10.29164/17charity

Bonney, R., & Hussain, A. (2001). *Faith communities and the development agenda*. Centre for the History of Religious and Political Pluralism, University of Leicester. http://www.dfid.gov.uk/pubs/files/faithdevcomagenda.pdf

Bruinessen, M. Van. (2019). Sufi "Orders" in Southeast Asia: From private devotions to social networks and corporate action. In R. M. Feener & A. M. Blackburn (Eds.), *Buddhist and Islamic orders in Southern Asia: Comparative perspectives* (1st ed., pp. 125–152). University of Hawai'i Press.

Candland, C. (2000). Faith as social capital: Religion and community development in Southern Asia. *Policy Sciences, 33*(3), 355–374.

Cannon, T., Twigg, J., & Rowell, J. (2003). *Social vulnerability, sustainable livelihoods and disasters*. DFID.

Chambers, R. (1984). *Rural development?* Longman.

Chambers, R. (1997). *Whose reality counts? Putting the first last*. IT Publications.

Chambers, R. (2002). Relaxed and participatory appraisal: Notes on practical approaches and methods for participants in PRA/PLA-related familiarisation workshops. In *Annals of the American Academy of Political and Social Science*. Institute for Development Studies (IDS), University of Sussex. http://www.idrc.ca/uploads/user-S/11491553671Reader_1_PRA_notes.pdf

Chambers, R. (2005). *Ideas for development*. Earthscan.

Chester, D. K., Duncan, A. M., & Dibben, C. J. L. (2008). The importance of religion in shaping volcanic risk perception in Italy, with special reference to Vesuvius and Etna. *Journal of Volcanology and Geothermal Research, 172*(3–4), 216–228. https://doi.org/10.1016/j.jvolgeores.2007.12.009

Chester, D. K., & Chester, O. K. (2010). The impact of eighteenth century earthquakes on the Algarve region, southern Portugal. *Geographical Journal, 176*(4), 350–370. https://doi.org/10.1111/j.1475-4959.2010.00367.x

Chester, D. K., & Duncan, A. M. (2010). Responding to disasters within the Christian tradition, with reference to volcanic eruptions and earthquakes. *Religion, 40*(2), 85–95. http://www.sciencedirect.com/science/article/B6WWN-4YBVN21-1/2/5c3bee10838a2891eb44345ec0641f24

Clarke, M. (2013). *Handbook of research on development and religion*. Edward Elgar Publishing Limited.

Crang, M., & Cook, I. (2007). *Doing ethnographies*. Sage.

Davis, N. J., & Robinson, R. V. (2012). *Claiming society for God: Religious movements and social welfare*. Indiana University Press.

De Cordier, B. (2010). On the thin line between good intentions and creating tensions: A view on gender programmes in Muslim contexts and the (potential) position of Islamic aid organisations. *European Journal of Development Research, 22*(2), 234–251. https://doi.org/10.1057/ejdr.2010.2

DeFranza, D., Lindow, M., Harrison, K., Mishra, A., & Mishra, H. (2020). Religion and reactance to COVID-19 mitigation guidelines. *American Psychologist*. https://doi.org/10.1037/amp0000717

Dekens, J. (2007). *Local knowledge for disaster preparedness: A literature review*. International Centre for Integrated Mountain Development (ICIMOD). http://books.icimod.org/index.php/search/publication/290

Deneulin, S., & Rakodi, C. (2011). Revisiting religion: Development studies thirty years on. *World Development, 39*(1), 45–54. https://doi.org/10.1016/j.worlddev.2010.05.007

Feener, R. M. (2013). *Shari'a and social engineering: The implementation of Islamic law in contemporary Aceh*. Oxford University Press.

Feener, R. M., & Fountain, P. (2018). Religion in the age of development. *Religions, 382*(9), 1–23. https://doi.org/10.1017/CBO9780511621475

Feener, R. M., & Scheer, C. (2018). Development's missions. In C. Scheer, P. Fountain, & R. M. Feener (Eds.), *The mission of development: Religion and techno-politics in Asia* (pp. 1–27). Brill.

Fountain, P., Bush, R., & Feener, R. M. (2015). Religion and the politics of development. In P. Fountain, R. Bush, & R. M. Feener (Eds.), *Religion and the politics of development* (pp. 11–34). Palgrave Macmillan.

Gaillard, J. C., & Texier, P. (2010). Religions, natural hazards, and disasters: An introduction. *Religion, 40*(2), 81–84. http://www.sciencedirect.com/

science/article/B6WWN-4YGHK8H-1/2/91b2bc41f3be1b58ac1ec07066f 66853

Gibson, M. (2006). *Order from chaos: Responding to traumatic events*. The Policy Press.

Gingerich, T. R., Moore, D. L., Brodrick, R., & Beriont, C. (2017). *Local humanitarian leadership and religious literacy: Engaging with religion, faith, and faith actors*. https://doi.org/10.21201/2017.9422

Haque, N. ul. (2009). How to solve Pakistan's problems. *Open Democracy*. http://www.opendemocracy.net/article/how-to-solve-pakistan-s-problem

Heijmans, A., Okechukwu, I., Peursum, A. S. tot, & Skarubowiz, R. (2009). A grassroots perspective on risks stemming from disasters and conflict. *Humanitarian Exchange, 44*, 34–35.

Hirschmann, D. (2003). Keeping "the last" in mind: Incorporating chambers in consulting. *Development in Practice, 13*(5), 487–500. https://doi.org/10.1080/0961452032000125866

Horton, L. (2012). After the earthquake: Gender inequality and transformation in post-disaster Haiti. *Gender & Development, 20*(2), 295–308. https://doi.org/10.1080/13552074.2012.693284

ICRC. (2020). *Pakistan: ICRC and Shariah Academy organize workshop on restraint in war*. https://www.icrc.org/en/document/pakistan-icrc-shariah-academy-workshop

IRRC. (2005). Editorial. *International Review of the Red Cross, 858*, 237–241. http://www.icrc.org/Web/Eng/siteeng0.nsf/html/review-858-p237

Ishiwatari, M. (2021). Institutional coordination of disaster management: Engaging national and local governments in Japan. *Natural Hazards Review, 22*(1), 04020059. https://doi.org/10.1061/(asce)nh.1527-6996.0000423

Islamic Ideology Council. (1993). *Social reformations* (in Urdu). Department of Computer and Publication.

Jamal, A. (2005). The political participation and engagement of Muslim Americans. *American Politics Research, 33*(4), 521–544. https://doi.org/10.1177/1532673x04271385

Kennedy, F. (2001). Sicilians pray as technology fails to stop lava. *The Independent*. http://www.encyclopedia.com/doc/1P2-5177590.html

Krull, L. M., Pearce, L. D., & Jennings, E. A. (2021). How religion, social class, and race intersect in the shaping of young women's understandings of sex, reproduction, and contraception. *Religions, 12*, 5.

Laditka, S. B., Murray, L. M., & Laditka, J. N. (2010). In the eye of the storm: Resilience and vulnerability among African American women in the wake of Hurricane Katrina. *Health Care for Women International, 31*(11), 1013–1027. https://doi.org/10.1080/07399332.2010.508294

Leftwich, A., & Sen, K. (2010). *Beyond institutions. Institutions and organizations in the politics and economics of poverty reduction - a thematic synthesis of research evidence*. DFID-funded Research Programme Consortium on improving Institutions for Pro-Poor Growth (IPPG).

McClure, J. (2017). Fatalism, causal reasoning, and natural hazards: Oxford research encyclopaedia of natural hazard science - oi. In *Oxford Research Encyclopedia of Natural Hazard Science*. https://doi.org/10.1093/acrefore/9780199389407.013.39

McGregor, A. (2010). Geographies of religion and development: Rebuilding sacred spaces in Aceh, Indonesia, after the tsunami. *Environment and Planning A, 42*(3), 729–746.

Ministry of Population Welfare. (2005). *Islamabad declaration: International Ulama conference on population and development March 4–6* (p. 7). Ministry of Population Welfare. http://www.mopw.gov.pk/ulamafollowup/declaration.pdf

Ni Raghallaigh, M., & Gilligan, R. (2010). Active survival in the lives of unaccompanied minors: Coping strategies, resilience, and the relevance of religion. *Child & Family Social Work, 15*(2), 226–237. https://doi.org/10.1111/j.1365-2206.2009.00663.x

Oktari, R. S., Kamaruzzaman, S., Fatimahsyam, F., Sofia, S., & Sari, D. K. (2021). Gender mainstreaming in a Disaster-Resilient Village Programme in Aceh Province, Indonesia: Towards disaster preparedness enhancement via an equal opportunity policy. *International Journal of Disaster Risk Reduction, 52*. https://doi.org/10.1016/j.ijdrr.2020.101974

Oliver-Smith, A., & Hoffman, S. M. (1999). *The angry earth: Disaster in anthropological perspective*. Routledge.

Oliver-Smith, A. (2004). Theorizing vulnerability in a globalized world: A political ecological perspective. In G. Bankoff, G. Frerks, & D. Hilhorst (Eds.), *Mapping vulnerability: Disaster, development and people* (pp. 1–10). Earthscan.

Paradise, T. R. (2005). Perception of earthquake risk in Agadir, Morocco: A case study from a Muslim community. *Global Environmental Change Part B: Environmental Hazards, 6*(3), 167–180. http://www.sciencedirect.com/science/article/B6VPC-4KPP4DD-1/2/4b72955cce74574bfb79d3b3e1c778c0

Pew Research Centre. (2012). *The global religious landscape*.

Rahmani, A. I. (2006). *The role of religious institutions in community governance affairs: How are communities governed beyond the district level?* Centre for Policy Studies, Central European University. http://pdc.ceu.hu/archive/00002849/01/rahmani.pdf

Reale, A. (2010). Acts of God(s): The role of religion in disaster risk reduction. *Humanitarian Exchange Magzine, 48*. http://www.odihpn.org/humanitarian-exchange-magazine/issue-48/acts-of-gods-the-role-of-religion-in-disaster-risk-reduction

Régnier, P., Neri, B., Scuteri, S., & Miniati, S. (2008). From emergency relief to livelihood recovery: Lessons learned from post-tsunami experiences in

Indonesia and India. *Disaster Prevention and Management: An International Journal, 17*(3), 410–429. https://doi.org/10.1108/09653560810887329

Rowel, R., Mercer, L. A., & Gichomo, G. (2011). Role of pastors in disasters curriculum development project: Preparing faith-based leaders to be agents of safety. *Journal of Homeland Security and Emergency Management, 8*(1), Article 32.

Seguino, S. (2011). Help or hindrance? Religion's impact on gender inequality in attitudes and outcomes. *World Development, 39*(8), 1308–1321. https://doi.org/10.1016/j.worlddev.2010.12.004

Sen, A. (1999). *Development as freedom*. Anchor Books.

Sheikhi, R. A., Seyedin, H., Qanizadeh, G., & Jahangiri, K. (2020). Role of religious institutions in disaster risk management: A systematic review. *Disaster Medicine and Public Health Preparedness*. https://doi.org/10.1017/dmp.2019.145

Simpson, E. (2014). *The political biography of an earthquake – aftermath and Amnesia in Gujarat, India (Issue 1)*. Oxford University Press.

Sudmeier-Rieux, K., Jaboyedoff, M., Breguet, A., Dubois, J., Peduzzi, P., Qureshi, R., & Jaubert, R. (2008). Strengthening decision-making tools for disaster risk reduction: An example of an integrative approach from Northern Pakistan. *IHDP Update: Magazine of the International Human Dimensions Programme on Global Environmental Change, 2*, 74–78.

Sudmeier-Rieux, K., Jaboyedoff, M., Breguet, A., & Dubois, J. (2011). The 2005 Pakistan earthquake revisited: Methods for integrated landslide assessment. *Mountain Research and Development, 31*(2), 112–121. https://doi.org/10.1659/mrd-journal-d-10-00110.1

Suri, K. (2018). Understanding historical, cultural and religious frameworks of mountain communities and disasters in Nubra valley of Ladakh. *International Journal of Disaster Risk Reduction, 31*, 504–513. https://doi.org/10.1016/J.IJDRR.2018.06.004

The Dawn. (2020, November 17). *President for engaging ulema in population control*. https://www.dawn.com/news/1590804/president-for-engaging-ulema-in-population-control

UNICEF. (2017). *Civil society partnerships: Framework for engagement*. https://www.unicef.org/about/partnerships/index_60134.html

UNICEF. (2020). Pakistan COVID-19 situation report no. 15. 15, 1–14. https://www.unicef.org/media/78396/file/Pakistan-COVID19-SitRep-15-August-2020.pdf. 15. Accessed 7 September 2020.

UNISDR. (2005). *Hyogo framework for action 2005–2015: Building the resilience of nations and communities to disasters*. In World Conference on Disaster Reduction, 18–22 January 2005, Kobe, Hyogo, Japan (Issue A/CONF.206/6). United Nations International Strategy for Disaster Reduction (UNISDR).

Vyborny, K. (2020). *Persuasion and public health: Evidence from an experiment with religious leaders during COVID-19 in Pakistan* (IGC Working Paper, 1–21). https://www.dropbox.com/s/eb04gvie3hp19kd/Imams_live_web.pdf?dl=0

Wisner, B. (1998, July 12–15). *World views, belief systems and disasters: Implications for preparedness, mitigation and recovery*. In Panel on world views and belief systems at the 23rd annual natural hazards research and application workshop, Boulder, Colorado.

Wisner, B. (2010). Untapped potential of the world's religious communities for disaster reduction in an age of accelerated climate change: An epilogue & prologue. *Religion, 40*(2), 128–131. http://www.sciencedirect.com/science/article/B6WWN-4YCNJPK-3/2/cf7680b3eda1132b8ff10f38a8690655

Bibliography

Abdel-Hady, Z. M. (2010). *The masjid, yesterday and today*. The Centre for International and Regional Studies, Georgetown University School of Foreign Service.

Akhuwat. (2021). *Akhuwat's story* (Akhuwat, Ed.). Akhuwat. http://www.akhuwat.org.pk

Al-Astewani, A. (2021). To open or close? COVID-19, mosques and the role of religious authority within the British Muslim community: A socio-legal analysis. *Religions, 12*(1), 1–26. https://doi.org/10.3390/rel12010011

Ali, I. (2008). Crises of governance. *International Institute for Asian Studies Newsletter, 49*(Autumn), 1–4. http://www.iias.nl/nl/49/IIAS_NL49_0104.pdf

Antonovsky, A. (1979). *Health, stress and coping*. Jossey-Bass Publishers.

Arutz Sheva. (2009). Joseph's era coins found in Egypt. *Arutz Sheva 7*. http://www.israelnationalnews.com/News/News.aspx/133601

Atallah, S., Khan, M. Z. A., & Malkawi, M. (2001). Water conservation through public awareness based on Islamic teachings in the Eastern Mediterranean region. In N. I. Faruqui, A. K. Biswas, & M. J. Bino (Eds.), *Water management in Islam* (pp. 49–60). United Nations University Press.

Atia, M. (2013). *Building a house in Heaven: Pious neoliberalism and Islamic charity in Egypt*. University of Minnesota Press.

Bankoff, G. (2004). The historical geography of disaster: Vulnerability and local knowledge. In G. Bankoff, G. Frerks, & D. Hilhorst (Eds.), *Mapping vulnerability: Disaster, development and people* (pp. 25–36). Earthscan.

A. R. Cheema, *The Role of Mosque in Building Resilient Communities*, Islam and Global Studies,
https://doi.org/10.1007/978-981-16-7600-0

Bano, M., & Nair, P. (2007). *Faith-based organizations in South Asia: Historical evolution, current status and nature of interaction with the state*. University of Birmingham.

BBC. (2008). *Clean water is essential for safeguarding eyes from blindness*. BBC URDU.Com. http://www.bbc.co.uk/urdu/multimedia/

BBC. (2021, January 21). *Birmingham mosque becomes UK's first to offer Covid vaccine*. https://www.bbc.com/news/uk-england-birmingham-55752056

Belshaw, D., Calderisi, R., & Sugden, C. (2001). *Faith in development: Partnership between the World Bank and the churches of Africa*. Regnum Books International.

Benthall, J. (2017). Charity. In F. Stein, S. Lazar, M. Candea, H. Diemberger, J. Robbinson, A. Sanchez, & R. Stasch (Eds.), *The Cambridge encyclopedia of anthropology* (pp. 1–15). Cambridge University Press. http://doi.org/10.29164/17charity

Berger, J. (2003). Religious nongovernmental organizations: An exploratory analysis. *Voluntas: International Journal of Voluntary and Nonprofit Organizations, 14*(1), 15–39.

Berkes, F. (2008). *Sacred ecology* (2nd ed.). Taylor & Francis.

Bernstein, P. L. (1996). *Against the Gods: The remarkable story of risk*. Wiley.

Bonney, R., & Hussain, A. (2001). *Faith communities and the development agenda*. Centre for the History of Religious and Political Pluralism, University of Leicester. http://www.dfid.gov.uk/pubs/files/faithdevcomagenda.pdf

Bruinessen, M. Van. (2019). Sufi "Orders" in Southeast Asia: From private devotions to social networks and corporate action. In R. M. Feener & A. M. Blackburn (Eds.), *Buddhist and Islamic orders in Southern Asia: Comparative perspectives* (1st ed., pp. 125–152). University of Hawai'i Press.

Brummitt, C. (2006). Indonesians see disasters as God's will. *The Christian Post*. http://www.christianpost.com/news/indonesians-see-disasters-as-god-s-will-1611/

Cairns, E. (2012). *Crises in a new world order: Challenging the humanitarian project*. Oxfam International.

Callisthenes. (1935). Our record of the Quetta earthquake. *The Times*, p. 10.

Candland, C. (2000). Faith as social capital: Religion and community development in Southern Asia. *Policy Sciences, 33*(3), 355–374.

Cannon, T., Twigg, J., & Rowell, J. (2003). *Social vulnerability, sustainable livelihoods and disasters*. DFID.

Chambers, R. (1984). *Rural development?* Longman.

Chambers, R. (1997). *Whose reality counts? Putting the first last*. IT Publications.

Chambers, R. (2002). Relaxed and participatory appraisal: Notes on practical approaches and methods for participants in PRA/PLA-related familiarisation workshops. In *Annals of the American Academy of Political and Social Science*.

Institute for Development Studies (IDS), University of Sussex. http://www.idrc.ca/uploads/user-S/11491553671Reader_1_PRA_notes.pdf

Chambers, R. (2005). *Ideas for development*. Earthscan.

Cheema, A. R., Mehmood, A., & Imran, M. (2016). Learning from the past: Analysis of disaster management structures, policies and institutions in Pakistan. *Disaster Prevention and Management, 25*(4), 449–463. https://doi.org/10.1108/DPM-10-2015-0243

Cheema, A. R., Scheyvens, R., Glavovic, B., & Imran, M. (2014). Unnoticed but important: Revealing the hidden contribution of community-based religious institution of the mosque in disasters. *Natural Hazards*, 1–23. https://doi.org/10.1007/s11069-013-1008-0

Cheema, A., Khwaja, A. I., & Qadir, A. (2006). Local government reforms in Pakistan: Context, content and causes. In P. Bardhan (Ed.), *Decentralization and local governance in developing countries* (pp. 381–433). MIT Press.

Cheema, A., & Mehmood, A. (2018). Reproductive health services: 'Business in a Box' as a model social innovation. *Development in Practice*. https://doi.org/10.1080/09614524.2018.1541166

Cheema, A., Khwaja, A. I., & Qadir, A. (2005). *Decentralization in Pakistan: Context, content and causes*. Kennedy School of Government, Harvard University, MA.

Chester, D. K. (2005). Theology and disaster studies: The need for dialogue. *Journal of Volcanology and Geothermal Research, 146*(4), 319–328. https://doi.org/10.1016/j.jvolgeores.2005.03.004

Chester, D. K., & Chester, O. K. (2010). The impact of eighteenth century earthquakes on the Algarve region, southern Portugal. *Geographical Journal, 176*(4), 350–370. https://doi.org/10.1111/j.1475-4959.2010.00367.x

Chester, D. K., & Duncan, A. M. (2010). Responding to disasters within the Christian tradition, with reference to volcanic eruptions and earthquakes. *Religion, 40*(2), 85–95. http://www.sciencedirect.com/science/article/B6WWN-4YBVN21-1/2/5c3bee10838a2891eb44345ec0641f24

Chester, D. K., Duncan, A. M., & Dibben, C. J. L. (2008). The importance of religion in shaping volcanic risk perception in Italy, with special reference to Vesuvius and Etna. *Journal of Volcanology and Geothermal Research, 172*(3–4), 216–228. https://doi.org/10.1016/j.jvolgeores.2007.12.009

Clarke, G. (2006). Faith matters: Faith-based organisations, civil society and international development. *Journal of International Development, 18*(6), 835–848. https://doi.org/10.1002/jid.1317

Clarke, G. (2007). Agents of transformation? Donors, faith-based organisations and international development. *Third World Quarterly, 28*(1), 77–96.

Clarke, M. (2013). *Handbook of research on development and religion*. Edward Elgar Publishing Limited.

Coppola, D. P. (2007). *Introduction to international disaster management*. Butterworth-Heinemann.

Cottrell, A. (2006). Weathering the storm: Women's preparedness as a form of resilience to weather related hazards in Northern Australia. In D. Paton & D. Johnston (Eds.), *Disaster resilience: An integrated approach* (pp. 128–142). Charles C Thomas.

Crang, M., & Cook, I. (2007). *Doing ethnographies*. Sage.

Davis, N. J., & Robinson, R. V. (2012). *Claiming society for God: Religious movements and social welfare*. Indiana University Press.

De Cordier, B. (2010). On the thin line between good intentions and creating tensions: A view on gender programmes in Muslim contexts and the (potential) position of Islamic aid organisations. *European Journal of Development Research, 22*(2), 234–251. https://doi.org/10.1057/ejdr.2010.2

DeFranza, D., Lindow, M., Harrison, K., Mishra, A., & Mishra, H. (2020). Religion and reactance to COVID-19 mitigation guidelines. *American Psychologist*. https://doi.org/10.1037/amp0000717

Dekens, J. (2007). *Local knowledge for disaster preparedness: A literature review*. International Centre for Integrated Mountain Development (ICIMOD). http://books.icimod.org/index.php/search/publication/290

Deneulin, S., & Zampini Davies, A. (2017). Engaging development and religion: Methodological groundings. *World Development, 99*, 110–121. https://doi.org/10.1016/J.WORLDDEV.2017.07.014

Deneulin, S., & Rakodi, C. (2011). Revisiting religion: Development studies thirty years on. *World Development, 39*(1), 45–54. https://doi.org/10.1016/j.worlddev.2010.05.007

Department for International Development. (2001). *Target 2015 halving world poverty: A shared vision of reducing world poverty – British Government and British Muslim charities working to realise the common good*. Department for International Development (DFID). http://www.dfid.gov.uk/pubs/files/2015-muslim.pdf

Dunn, K. M. (2001). Representations of Islam in the politics of mosque development in Sydney. *Tijdschrift Voor Economische En Sociale Geografie, 92*(3), 291–308. https://doi.org/10.1111/1467-9663.00158

Earthquake Reconstruction and Rehabilitation Authority. (2005). *Earthquake Reconstruction and Rehabilitation Authority (ERRA) ordinance* (G. of P. Earthquake Reconstruction and Rehabilitation Authority (ERRA), Ed.). http://www.erra.pk/aboutus/erra.asp#Ordinance

Earthquake Reconstruction and Rehabilitation Authority. (2009). *Information booklet for Union Council Disaster Management Committee (UCDMC)*. Earthquake Reconstruction and Rehabilitation Authority (ERRA), Government of Pakistan.

Earthquake Reconstruction and Rehabilitation Authority. (2010a). Community *based disaster risk management (CBDRM)* (E. R. and R. Authority, Ed.; Issue 2010). Earthquake Reconstruction and Rehabilitation Authority, Government of Pakistan. http://www.erra.pk/sectors/drr/CBDRM.asp

Earthquake Reconstruction and Rehabilitation Authority. (2010b). *Earthquake Reconstruction and Rehabilitation Authority (ERRA) - About us*. http://www.erra.pk/aboutus/erra.asp#HB

Earthquake Reconstruction and Rehabilitation Authority. (2011). *Annual review 2009–2010*. Earthquake Reconstruction and Rehabilitation Authority (ERRA).

Federal Emergency Management Association. (2010a). *Mitigation*. Federal Emergency Management Association (FEMA).

Federal Emergency Management Association. (2010b). *Mitigation best practices portfolio*. http://www.fema.gov/plan/prevent/bestpractices/index.shtm

Federal Emergency Management Association. (2011). *FEMA mitigation and insurance plan*. Federal Emergency Management Association (FEMA).

Feener, R. M. (2013). *Shari'a and social engineering: The implementation of Islamic law in contemporary Aceh*. Oxford University Press.

Feener, R. M., & Fountain, P. (2018). Religion in the age of development. *Religions, 382*(9), 1–23. https://doi.org/10.1017/CBO9780511621475

Feener, R. M., & Scheer, C. (2018). Development's missions. In C. Scheer, P. Fountain, & R. M. Feener (Eds.), *The mission of development: Religion and techno-politics in Asia* (pp. 1–27). Brill.

Flood Forecasting Division. (2005, January 18–22). Summary of national information on the current status of disaster reduction, as background for the World Conference on Disaster Reduction (WCDR). Pakistan: Vol. January. https://www.unisdr.org/2005/wcdr/preparatory-process/national-reports/summary-national-reports.pdf

Fountain, P., Bush, R., & Feener, R. M. (2015). Religion and the politics of development. In P. Fountain, R. Bush, & R. M. Feener (Eds.), *Religion and the politics of development* (pp. 11–34). Palgrave Macmillan.

Fozdar, F., & Roberts, K. (2010). Islam for fire fighters - a case study on an education programme for emergency services. *The Australian Journal of Emergency Management, 25*(1), 47–53. http://www.ema.gov.au/www/emaweb/rwpattach.nsf/VAP/%288AB0BDE05570AAD0EF9C283AA8F533E3%29~Roberts+&+Fozdar.pdf/$file/Roberts+&+Fozdar.pdf

Gaillard, J. C., & Texier, P. (2010). Religions, natural hazards, and disasters: An introduction. *Religion, 40*(2), 81–84. http://www.sciencedirect.com/science/article/B6WWN-4YGHK8H-1/2/91b2bc41f3be1b58ac1ec07066f66853

Gianisa, A., & Le De, L. (2018). The role of religious beliefs and practices in disaster: The case study of 2009 earthquake in Padang city, Indonesia.

Disaster Prevention and Management: An International Journal, 27(1), 74–86. https://doi.org/10.1108/DPM-10-2017-0238

Gibson, M. (2006). *Order from chaos: Responding to traumatic events*. The Policy Press.

Gingerich, T. R., Moore, D. L., Brodrick, R., & Beriont, C. (2017). *Local humanitarian leadership and religious literacy: Engaging with religion, faith, and faith actors*. https://doi.org/10.21201/2017.9422

Disaster Management Act 2010. (2011). http://web.ndma.gov.pk/files/NDMA-Act.pdf

Government of Pakistan. (2011). *National Volunteer Movement (NVM): About us*. http://www.nvm.org.pk/AboutUs/Index.html

Government of the Punjab. (2019). *Provincial disaster management authority: DRM institutions*. http://pdma.gop.pk/drm_institutions

Government of the United States. (1988). *Robert T. Stafford disaster relief and emergency assistance Act (Public Law 93–288) as amended upto June 2007* (F. E. M. A. (FEMA), Ed.). http://www.fema.gov/about/stafact.shtm

Grandjean, D., Rendu, A.-C., MacNamee, T., & Scherer, K. R. (2008). The wrath of the gods: Appraising the meaning of disaster. *Social Science Information, 47*(2), 187–204. https://doi.org/10.1177/0539018408089078

Grare, F. (2007). The evolution of sectarian conflicts in Pakistan and the ever-changing face of Islamic violence. *South Asia-Journal of South Asian Studies, 30*(1), 127–143. https://doi.org/10.1080/00856400701264068

Griffiths, V. F., Bull, J. W., Baker, J., Infield, M., Roe, D., Nalwanga, D., Byaruhanga, A., & Milner-Gulland, E. J. (2020). Incorporating local nature-based cultural values into biodiversity No Net Loss strategies. *World Development, 128*, 104858. https://doi.org/10.1016/j.worlddev.2019.104858

Guarnacci, U. (2016). Joining the dots: Social networks and community resilience in post-conflict, post-disaster Indonesia. *International Journal of Disaster Risk Reduction, 16*, 180–191. https://doi.org/10.1016/j.ijdrr.2016.03.001

Gupta, D. K., & Mundra, K. (2005). Suicide bombing as a strategic weapon: An empirical investigation of Hamas and Islamic Jihad. *Terrorism and Political Violence, 17*(4), 573–598. https://doi.org/10.1080/09546550500189895

Ha, K. M. (2015). The role of religious beliefs and institutions in disaster management: A case study. *Religions, 6*(4), 1314–1329. https://doi.org/10.3390/rel6041314

Haddad, Y. Y., & Balz, M. J. (2008). Taming the Imams: European governments and Islamic preachers since 9/11. *Islam and Christian–Muslim Relations, 19*(2), 215–235. http://www.informaworld.com/10.1080/09596410801923980

Haleem, I. (2003). Ethnic and sectarian violence and the propensity towards praetorianism in Pakistan. *Third World Quarterly, 24*(3), 463–477. https://doi.org/10.1080/0143659032000084410

Haqqani, H. (2005). *Pakistan: Between mosque and military*. United Book Press.

Haque, N. ul. (2009). How to solve Pakistan's problems. *Open Democracy*. http://www.opendemocracy.net/article/how-to-solve-pakistan-s-problem

Harrison, T. G. (1992). *Peacetime employment of the military: The Army's role in domestic disaster relief*. In U.S Army War College, Carlisle Barracks. U.S Army War College, Carlisle Barracks.

Heijmans, A., Okechukwu, I., Peursum, A. S. tot, & Skarubowiz, R. (2009). A grassroots perspective on risks stemming from disasters and conflict. *Humanitarian Exchange, 44*, 34–35.

Hirono, T., & Blake, M. E. (2017). The role of religious leaders in the restoration of hope following natural disasters. *SAGE Open, 7*(2). https://doi.org/10.1177/2158244017707003

Hirschmann, D. (2003). Keeping "the last" in mind: Incorporating chambers in consulting. *Development in Practice, 13*(5), 487–500. https://doi.org/10.1080/0961452032000125866

Holmgaard, S. B. (2019). The role of religion in local perceptions of disasters: The case of post-tsunami religious and social change in Samoa. *Environmental Hazards, 18*(4), 311–325. https://doi.org/10.1080/17477891.2018.1546664

Homan, J. (2003). The social construction of natural disaster: Egypt and the UK. In M. Pelling (Ed.), *Natural disasters and development in a globalizing world* (pp. 141–156). Routledge.

Horton, L. (2012). After the earthquake: Gender inequality and transformation in post-disaster Haiti. *Gender & Development, 20*(2), 295–308. https://doi.org/10.1080/13552074.2012.693284

Hutton, D., & Haque, C. E. (2003). Patterns of coping and adaptation among erosion-induced displacees in Bangladesh: Implications for hazard analysis and mitigation. *Natural Hazards, 29*(3), 405–421. https://doi.org/10.1023/A:1024723228041

ICRC. (2020). *Pakistan: ICRC and Shariah Academy organize workshop on restraint in war*. https://www.icrc.org/en/document/pakistan-icrc-shariah-academy-workshop

Imran, M. (2010). Sustainable urban transport in Pakistan: An institutional analysis. *International Planning Studies, 15*(2), 119–141.

Innes, M. (2006). Policing uncertainty: Countering terror through community intelligence and democratic policing. *The ANNALS of the American Academy of Political and Social Science, 605*(1), 222.

IRRC. (2005). Editorial. *International Review of the Red Cross, 858*, 237–241. http://www.icrc.org/Web/Eng/siteeng0.nsf/html/review-858-p237

Ishiwatari, M. (2021). Institutional coordination of disaster management: Engaging national and local governments in Japan. *Natural Hazards Review, 22*(1), 04020059. https://doi.org/10.1061/(asce)nh.1527-6996.0000423

Islamic Ideology Council. (1993). *Social reformations* (in Urdu). Department of Computer and Publication.

Jacob, T. (2001). Path-dependent Danish welfare reforms: The contribution of the new institutionalisms to understanding evolutionary change. *Scandinavian Political Studies, 24*(4), 277–309.

Jamal, A. (2005). The political participation and engagement of Muslim Americans. *American Politics Research, 33*(4), 521–544. https://doi.org/10.1177/1532673x04271385

Joakim, E. P., & White, R. S. (2015). Exploring the impact of religious beliefs, leadership, and networks on response and recovery of disaster-affected populations: A case study from Indonesia. *Journal of Contemporary Religion, 30*(2). https://doi.org/10.1080/13537903.2015.1025538

Kennedy, F. (2001). Sicilians pray as technology fails to stop lava. *The Independent*. http://www.encyclopedia.com/doc/1P2-5177590.html

Khan, W. (2011). *Say what we should do?* BBC Urdu.Com. http://www.bbc.co.uk/urdu/pakistan/2011/09/110928_naukot_wusat_ar.shtml

Krull, L. M., Pearce, L. D., & Jennings, E. A. (2021). How religion, social class, and race intersect in the shaping of young women's understandings of sex, reproduction, and contraception. *Religions, 12*, 5.

Laditka, S. B., Murray, L. M., & Laditka, J. N. (2010). In the eye of the storm: Resilience and vulnerability among African American women in the wake of Hurricane Katrina. *Health Care for Women International, 31*(11), 1013–1027. https://doi.org/10.1080/07399332.2010.508294

Leach, M., MacGregor, H., Scoones, I., & Wilkinson, A. (2020). Post-pandemic transformations: How and why COVID-19 requires us to rethink development. *World Development*, 105233. https://doi.org/10.1016/j.worlddev.2020.105233

Leftwich, A., & Sen, K. (2010). *Beyond institutions. Institutions and organizations in the politics and economics of poverty reduction - a thematic synthesis of research evidence*. DFID-funded Research Programme Consortium on improving Institutions for Pro-Poor Growth (IPPG).

Mahr, M. A. D. (2005). *The role of the mosque in the building of the society* (in Urdu). An-Noor Publications.

Malthus, T. R. (1958). *An essay on population* (M. P. Fogarty, Ed., Vol. 1). Dent.

Martens, K. (2002). Mission impossible? Defining nongovernmental organizations. *Voluntas: International Journal of Voluntary and Non-Profit Organizations, 13*(3), 271–285.

McClure, J. (2017). Fatalism, causal reasoning, and natural hazards: Oxford research encyclopaedia of natural hazard science - oi. In *Oxford Research Encyclopedia of Natural Hazard Science*. https://doi.org/10.1093/acrefore/9780199389407.013.39

McEntire, D. A. (2007). *Disaster response and recovery: Strategies and tactics for resilience*. Wiley.

McGeehan, K. M., & Baker, C. K. (2017). Religious narratives and their implications for disaster risk reduction. *Disasters, 41*(2), 258–281. https://doi.org/10.1111/disa.12200

McGregor, A. (2010). Geographies of religion and development: Rebuilding sacred spaces in Aceh, Indonesia, after the tsunami. *Environment and Planning A, 42*(3), 729–746.

Ministry of Civil Defence & Emergency Management. (2005). *Focus on recovery: A holistic framework for recovery in New Zealand* (M. of C. D. & E. Management, Ed.). Ministry of Civil Defence & Emergency Management (MCDEM), Government of New Zealand. http://www.civildefence.govt.nz

Ministry of Civil Defence & Emergency Management, & Affairs, D. of I. (2007). *National civil defence emergency management strategy*. Ministry of Civil Defence & Emergency Management (MCDEM), Government of New Zealand.

Ministry of Population Welfare. (2005). *Islamabad declaration: International Ulama conference on population and development March 4–6* (p. 7). Ministry of Population Welfare. http://www.mopw.gov.pk/ulamafollowup/declaration.pdf

Ministry of Population Welfare. (2006). *Proceedings of follow-up meeting of the council of participating countries: Islamabad declaration on population and development* (M. of P. Welfare, Ed.). http://www.mopw.gov.pk/ulamafollowup/fupindex.html

National Disaster Management Authority. (2007). *National disaster management framework of Pakistan*. National Disaster Management Authority (NDMA) Government of Pakistan.

National Disaster Management Authority. (2019). *National disaster response plan* (Issue 1). http://cms.ndma.gov.pk/

National Disaster Management Authority. (2021). *National Disaster Management Authority (NDMA): About us*. http://cms.ndma.gov.pk/page/about-us

Neal, D. M. (1997). Reconsidering the phases of disaster. *International Journal of Mass Emergencies and Disasters, 15*(2), 239–264.

Ngin, C., Grayman, J. H., Neef, A., & Sanunsilp, N. (2020). The role of faith-based institutions in urban disaster risk reduction for immigrant communities. *Natural Hazards, 103*(1), 299–316. https://doi.org/10.1007/s11069-020-03988-9

Ni Raghallaigh, M., & Gilligan, R. (2010). Active survival in the lives of unaccompanied minors: Coping strategies, resilience, and the relevance of religion. *Child & Family Social Work, 15*(2), 226–237. https://doi.org/10.1111/j.1365-2206.2009.00663.x

North, D. (1991). Institutions. *Journal of Economic Perspectives, 5*(1), 97–112.

Oktari, R. S., Kamaruzzaman, S., Fatimahsyam, F., Sofia, S., & Sari, D. K. (2021). Gender mainstreaming in a Disaster-Resilient Village Programme in Aceh Province, Indonesia: Towards disaster preparedness enhancement via an equal opportunity policy. *International Journal of Disaster Risk Reduction, 52*. https://doi.org/10.1016/j.ijdrr.2020.101974

Oliver-Smith, A., & Hoffman, S. M. (1999). *The angry earth: Disaster in anthropological perspective*. Routledge.

Oliver-Smith, A. (2004). Theorizing vulnerability in a globalized world: A political ecological perspective. In G. Bankoff, G. Frerks, & D. Hilhorst (Eds.), *Mapping vulnerability: Disaster, development and people* (pp. 1–10). Earthscan.

Paradise, T. R. (2005). Perception of earthquake risk in Agadir, Morocco: A case study from a Muslim community. *Global Environmental Change Part B: Environmental Hazards, 6*(3), 167–180. http://www.sciencedirect.com/science/article/B6VPC-4KPP4DD-1/2/4b72955cce74574bfb79d3b3e1c778c0

Pew Research Centre. (2012). *The global religious landscape*.

Planning Commission of Pakistan. (2010). *Government of Pakistan five year plans; Ist plan 1955–60, 2nd plan 1960–65, 3rd plan 1965–70, no plan 1971–76, 5th plan 1977–83, 6th plan 1983–88, 7th plan 1988–93, 8th plan 1993–98 and medium term development plan 2005–10*. http://www.planningcommission.gov.pk/

Platt, R. H. (1998). Planning and land use adjustments in historical perspective. In R. J. Burby (Ed.), *Cooperating with nature: Confronting natural hazards with land-use planning for sustainable communities* (pp. 29–56). Joseph Henry.

Platteau, J. P. (2010). Political instrumentalization of Islam and the risk of obscurantist deadlock. *World Development, 39*(2), 243–260.

Pooley, J. A., Cohen, L., & O'Connor, M. (2006). Links between community and individual resilience: Evidence from cyclone affected communities in North West Australia. In D. Paton & D. Johnston (Eds.), *Disaster resilience: An integrated approach* (pp. 161–173). Charles C Thomas.

Provincial Disaster Management Authority Punjab. (2018). *Disaster risk reduction strategy: Provincial disaster response plan*. http://pdma.gop.pk/system/files/DisasterRiskReductionStrategy-ProvincialDisasterResponsePlan2018%28Final%29_0.pdf

Quarantelli, E. L. (2009). *The earliest interest in disasters and crises and the early social science studies of disasters as seen in a sociology of knowledge perspective* (Working Paper 91). Disaster Research Center, University of Delaware.

Quarantelli, E. L., Lagadec, P., & Boin, A. (2007). A heuristic approach to future disasters and crises: New, old, and in-between types. In H. Rodriguez, E. L. Quarantelli, & R. R. Dynes (Eds.), *Handbooks of disaster research* (pp. 16–41). Springer.

Qureshi, J. (2006). Earthquake jihad: The role of jihadis and Islamist groups after the October 2005 earthquake. *Humanitarian Exchange, June*(34), 40. https://odihpn.org/wp-content/uploads/2006/07/humanitarianexchange034.pdf

Rahman, A., Shaw, R., & Khan, A. N. (2015). *Disaster risk reduction approaches in Pakistan*. Springer.

Rahmani, A. I. (2006). *The role of religious institutions in community governance affairs: How are communities governed beyond the district level?* Centre for Policy Studies, Central European University. http://pdc.ceu.hu/archive/00002849/01/rahmani.pdf

Reale, A. (2010). Acts of God(s): The role of religion in disaster risk reduction. *Humanitarian Exchange Magzine, 48*. http://www.odihpn.org/humanitarian-exchange-magazine/issue-48/acts-of-gods-the-role-of-religion-in-disaster-risk-reduction

Régnier, P., Neri, B., Scuteri, S., & Miniati, S. (2008). From emergency relief to livelihood recovery: Lessons learned from post-tsunami experiences in Indonesia and India. *Disaster Prevention and Management: An International Journal, 17*(3), 410–429. https://doi.org/10.1108/09653560810887329

Ronan, K. R., & Johnston, D. M. (2005). *Promoting community resilience in disasters: The role for schools, youth, and families*. Springer.

Rowel, R., Mercer, L. A., & Gichomo, G. (2011). Role of pastors in disasters curriculum development project: Preparing faith-based leaders to be agents of safety. *Journal of Homeland Security and Emergency Management, 8*(1), Article 32.

Sain, V. (2008). Jodhpur Muslims help temple stampede victims. *Hindustan Times*. http://www.hindustantimes.com/StoryPage/FullcoverageStoryPage.aspx?id=0bbee88b-fff7-424e-be94-479a78b51c5bTemplesofdoom_Special&&Headline=Jodhpur+Muslims+help+temple+stampede+victims

Schmuck, H. (2000). " An act of Allah": Religious explanations for floods in Bangladesh as survival strategy. *International Journal of Mass Emergencies and Disasters, 18*(1), 85–96.

Seguino, S. (2011). Help or hindrance? Religion's impact on gender inequality in attitudes and outcomes. *World Development, 39*(8), 1308–1321. https://doi.org/10.1016/j.worlddev.2010.12.004

Sen, A. (1999). *Development as freedom*. Anchor Books.

Shah, I., Eali, N., Alam, A., Dawar, S., & Dogar, A. A. (2020). Institutional arrangement for disaster risk management: Evidence from Pakistan. *International Journal of Disaster Risk Reduction, 51*(August), 101784. https://doi.org/10.1016/j.ijdrr.2020.101784

Shah, S. M. S., Baig, M. A., Khan, A. A., & Malkawi, M. (2001). Water conservation through community institutions in Pakistan: Mosques and religious schools. In N. I. Faruqui, A. K. Biswas, & M. J. Bino (Eds.), *Water management in Islam* (pp. 61–67). United Nations University Press.

Shaheedi, A. (2021, March). *GB provincial status*. The News International. https://www.thenews.com.pk/print/805077-gb-provincial-status

Sheikhi, R. A., Seyedin, H., Qanizadeh, G., & Jahangiri, K. (2020). Role of religious institutions in disaster risk management: A systematic review. *Disaster Medicine and Public Health Preparedness*. https://doi.org/10.1017/dmp.2019.145

Shore, B. (2002). Taking culture seriously. *Human Development, 45*(4), 228–266.

Siddiqa, A. (2007). *Military Inc: Inside Pakistan's military economy*. Oxford.

Sightsavers. (2021). *Our work in Pakistan*. http://www.sightsavers.org/whatwedo/ourworkintheworld/asia/pakistan/world6206.html

Simpson, E. (2014). *The political biography of an earthquake – aftermath and Amnesia in Gujarat, India (Issue 1)*. Oxford University Press.

Skrine, C. P. (1936). The Quetta earthquake. *The Geographical Journal, 88*(5), 414–428. http://www.jstor.org/stable/1785962

Socio-engineering Consultants. (2006). *Capacity building handbook for school management committee: Earthquake emergency assistance programme (EAAP)*. Socio-engineering Consultants.

Stern, G. (2007). *Can God intervene?: How religion explains natural disasters*. Praeger Publishers.

Sudmeier-Rieux, K., Jaboyedoff, M., Breguet, A., Dubois, J., Peduzzi, P., Qureshi, R., & Jaubert, R. (2008). Strengthening decision-making tools for disaster risk reduction: An example of an integrative approach from Northern Pakistan. *IHDP Update: Magazine of the International Human Dimensions Programme on Global Environmental Change, 2*, 74–78.

Sudmeier-Rieux, K., Jaboyedoff, M., Breguet, A., & Dubois, J. (2011). The 2005 Pakistan earthquake revisited: Methods for integrated landslide assessment. *Mountain Research and Development, 31*(2), 112–121. https://doi.org/10.1659/mrd-journal-d-10-00110.1

Suhail, R. (2010). *Temples are better than camps*. BBC Urdu.Com. http://www.bbc.co.uk/urdu/pakistan/2010/09/100902_flood_hindu_temple.shtml

Suri, K. (2018). Understanding historical, cultural and religious frameworks of mountain communities and disasters in Nubra valley of Ladakh. *International*

Journal of Disaster Risk Reduction, 31, 504–513. https://doi.org/10.1016/J.IJDRR.2018.06.004

The Dawn. (2009). Rescue-1122 launched in Peshawar. *The Dawn*. http://www.dawn.com/wps/wcm/connect/dawn-content-library/dawn/the-newspaper/national/rescue1122-launched-in-peshawar-189

The Dawn. (2020, November 17). President for engaging ulema in population control. https://www.dawn.com/news/1590804/president-for-engaging-ulema-in-population-control

The Express Tribune. (2019, November 8). *ERRA to be subsumed into NDMA*. https://tribune.com.pk/story/2095704/1-erra-subsumed-ndma-dec-31

The Times. (1935a). The earthquake at Quetta: General's praise of the garrison. *The Times*, p. 4.

The Times. (1935b). The earthquake at Quetta. *The Times*, p. 15.

Thompson, W. C. (2010). Success in Kashmir: A positive trend in civil–military integration during humanitarian assistance operations. *Disasters, 34*(1), 1–15. https://doi.org/10.1111/j.1467-7717.2009.01111.x

Thomson, J. (1995). *Community institutions and the governance of local woodstocks in the context of Mali's democratic transition*. 38th Annual Meeting of the African Studies Association, November 3–6. http://dlc.dlib.indiana.edu/archive/00002673/01/Community_Institutions.pdf

Times of India. (2009). 1.6 million Pakistani refugees return home: UN. *Times of India*. http://articles.timesofindia.indiatimes.com/2009-08-22/pakistan/28191405_1_refugees-return-home

Twigg, J. (2004). Good practice review: Disaster risk reduction, mitigation and preparedness in development and emergency programming. In *Disaster risk reduction: Mitigation and preparedness in development and emergency programming*. Overseas Development Institute. http://www.cababstractsplus.org/google/abstract.asp?AcNo=20043073047

UNICEF. (2017). *Civil society partnerships: Framework for engagement*. https://www.unicef.org/about/partnerships/index_60134.html

UNICEF. (2020). Pakistan COVID-19 Situation Report No. 15. 15, 1–14. https://www.unicef.org/media/78396/file/Pakistan-COVID19-SitRep-15-August-2020.pdf. 15. Accessed 7 September 2020.

UNISDR. (2005a). *Hyogo framework for action 2005–2015: Building the resilience of nations and communities to disasters*. In World Conference on Disaster Reduction, 18–22 January 2005, Kobe, Hyogo, Japan (Issue A/CONF.206/6). United Nations International Strategy for Disaster Reduction (UNISDR).

UNISDR. (2005b). *World conference on disaster reduction: 18–22 January*. United Nations International Strategy for Disaster Reduction (UNISDR). http://www.unisdr.org/eng/hfa/docs/Hyogo-framework-for-action-english.pdf

UNISDR. (2009a). *UNISDR terminology on disaster risk reduction*. United Nations International Strategy for Disaster Reduction (UNISDR). http://www.unisdr.org/eng/library/UNISDR-terminology-2009-eng.pdf

UNISDR. (2009b). 2009 UNISDR terminology on disaster risk reduction. United Nations International Strategy for Disaster Reduction (UNISDR), United Nations Office for Disaster Risk Reduction. https://www.undrr.org/publication/2009-unisdr-terminology-disaster-risk-reduction. Accessed 24 November 2021.

UNISDR. (2015). *Sendai framework for disaster risk reduction*. UNISDR. https://www.preventionweb.net/files/43291_sendaiframeworkfordrren.pdf

United Nations. (2018). *2018 Review of SDGs implementation: SDG 11 – Make cities and human settlements inclusive, safe, resilient and sustainable*. https://sustainabledevelopment.un.org/content/documents/197282018_background_notes_SDG_11_v3.pdf

Usmani, Z. U., Imana, E. Y., & Kirk, D. (2010). Escaping death - Geometrical recommendations for high value targets. In T. Sobh (Ed.), *Innovations and advances in computer sciences and engineering* (pp. 503–508). Springer-Verlag Berlin. https://doi.org/10.1007/978-90-481-3658-2_88

Vaught, S. (2009). An "Act of God": Race, religion, and policy in the wake of Hurricane Katrina. *Souls, 11*(4), 408–421. https://doi.org/10.1080/10999940903417276

Vyborny, K. (2020). *Persuasion and public health: Evidence from an experiment with religious leaders during COVID-19 in Pakistan* (IGC Working Paper, 1–21). https://www.dropbox.com/s/eb04gvie3hp19kd/Imams_live_web.pdf?dl=0

Wilkinson, O., Duff, J., Nam, S., Trotta, S., & Goodwin, E. (2020). *COVID 19: Practising our faith safely during a Pandemic - Adapting how we gather together, pray and practise*. Faith and Positive Change for Children, Families and Communities (FPCC). https://jliflc.com/covid/

Wisner, B. (1998, July 12–15). *World views, belief systems and disasters: Implications for preparedness, mitigation and recovery*. In Panel on world views and belief systems at the 23rd annual natural hazards research and application workshop, Boulder, Colorado.

Wisner, B. (2010). Untapped potential of the world's religious communities for disaster reduction in an age of accelerated climate change: An epilogue & prologue. *Religion, 40*(2), 128–131. http://www.sciencedirect.com/science/article/B6WWN-4YCNJPK-3/2/cf7680b3eda1132b8ff10f38a8690655

Wong, L. P. (2010). Information needs, preferred educational messages and channel of delivery, and opinion on strategies to promote organ donation: A multicultural perspective. *Singapore Medical Journal, 51*(10), 790–795.

Index

A. R. Cheema, *The Role of Mosque in Building Resilient Communities*, Islam and Global Studies,
https://doi.org/10.1007/978-981-16-7600-0

U

V

W

Z